GEOLOGY

GEOLOGY

AN ILLUSTRATED HISTORY

DAVID BAINBRIDGE

PRINCETON UNIVERSITY PRESS
PRINCETON & OXFORD

Published in 2026 by
Princeton University Press
41 William Street
Princeton, New Jersey 08540

99 Banbury Road,
Oxford OX2 6JX

press.princeton.edu

GPSR Authorized Representative: Easy Access System
Europe - Mustamäe tee 50, 10621 Tallinn, Estonia, gpsr.
requests@easproject.com

ISBN: 978-0-691-26983-2
Ebook ISBN: 978-0-691-27309-9

Library of Congress Control Number: 2025936873

British Library Cataloging-in-Publication Data
is available.

Conceived, edited and designed by
Quintessence Editions, an imprint of
The Quarto Group
1 Triptych Place, Second Floor
London SE1 9SH
www.quarto.com

QUAR.1179110

For Quintessence:
Senior Commissioning Editor: Eszter Karpati
Senior Editor: Kath Stathers
Copyeditor: Blanche Craig
Design: Blok Graphic
Senior Designer: Rachel Cross
Picture Research: Susannah Jayes
Production Manager: David Hearn
Managing Editor: Emma Harverson
Art Director: Gemma Wilson
Publisher, Quintessence: Eszter Karpati
Publisher: Lorraine Dickey

Cover image: Aspen Mountain Sheet. Sheet XXVI Sections.
U. S. Geological Survey, Charles D. Walcott, Director.
Monograph XXXI. Julius Bien & Co. Lith. N.Y. (1898).
Courtesy of the David Rumsey Map Collection,
David Rumsey Map Center, Stanford Libraries.

Cover design: Wanda España

Printed in Malaysia

10 9 8 7 6 5 4 3 2 1

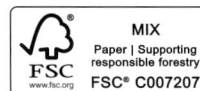

MIX
Paper | Supporting
responsible forestry
FSC® C007207

Contents

Introduction   6

1 TIME   12

2 ENERGY   56

3 PROCESS   106

4 USE   156

5 LIFE   202

Index   250

Image credits   256

> "There will always be rocks in the road ahead of us. They will be stumbling blocks or stepping stones; it all depends on how you use them."
>
> Attr. Friedrich Nietzsche (1844–1900)

Stepping Stones

The Earth is our familiar home, yet the geological processes which underlie our tenure on it can seem intimidatingly vast, ancient, and sometimes even alien—certainly beyond everyday human experience. Our planet's origins and upheavals have fascinated humans for millennia, but only recently have we developed a clear understanding of how the Earth actually works.

Many people alive today were born at a time when no one knew where Earth's constituents came from, why continents move, how diamonds are made, what causes earthquakes, when life started, or whether geological processes occur elsewhere in the Universe. We now know a great deal about all these things, and geology is expanding at breakneck speed.

Long before it had a name, geology was a crucial part of human knowledge. Even before agriculture, humans needed to know where water was wrung from the ground, where the best tool-stones could be found, and how to choose a stable, safe campsite. This knowledge expanded with the advent of farming and settlements around ten thousand years ago—working the land where minerals in the soil would make crops and beasts thrive, carving irrigation channels and trackways into the ground, and hewing stones to build the new villages and cities.

However, true geology (in Greek *ge* and *logos* translate as "Earth-reason") is more than just a corpus of factual knowledge that allows us to exploit the Earth: it is also a central part of our enlightened quest to comprehend our world. Progress in geology has often come from reinterpretation of geological phenomena which had always been visible to us, rather than the discovery of new sites and formations. All that was needed was

Leonardo da Vinci (1452–1519), *A Ravine with Waterbirds*, c. 1482–85.

As well as his artistic obsession with natural landforms, Leonardo made major contributions to early geology, examining the natural phenomena around him, and drawing conclusions which underpin our modern concepts of geological time, energy, and life.

thinkers with the insight to look afresh at what was always there. And the timescales of the young science of geology have been unusual: near-non-existent 250 years ago, and only finding its central concept—plate tectonics—in the 1960s.

People even argue about what geology encompasses. Increasingly the term "Earth sciences" is used, as it is more flexible and allows the discipline to spill over into study of the Earth's oceans and atmosphere. Also, confusingly, we now use "geology" to apply to processes taking place on other planets and moons—as far away from *ge* as one can get. In this book, I have tried to restrict myself to geology as "solid Earth science"—the hard, dense stuff. However, it is impossible to be entirely consistent because the hard components of the planet are in continual interaction with its watery, icy, and gaseous coatings. Also, some of the Earth's dense parts are themselves liquid anyway, or liquid-like at least.

Geology started out promisingly. As in most areas of science it was the ancient Greeks who first framed the right questions, even if they did not have enough information at that time to work their way toward the right answers.

The Greek island of Thira, or Santorini, 2018.

Santorini is the remains of a violent volcanic eruption which destroyed coastal towns across the eastern Mediterranean, changing the course of Bronze Age history.

As early as the eighth century BCE the epic poet Hesiod attempted to explain the vast forces driving Earth's processes. The cataclysmic explosion of the Aegean island of Thira eight hundred years earlier had devastated Bronze Age settlements across the eastern Mediterranean, and cast a long cultural pall over the region. Hesiod implied parallels between such volcanic events and mythic battles between Zeus and dragons and Titans. Later, Pythagoras (c. 570–c. 495 BCE) referred to eternal interconversion of the different constituents of the Earth, Plato (427–348 BCE) hypothesized rivers of fire within the earth, and Aristotle (384–322 BCE) pondered erosion, silting, and the creation of different minerals.

Among the ancient Greeks' many achievements, the most remarkable one was by the less famous Eratosthenes (c. 276–c. 194 BCE) who very much *did* get the right answer. Eratosthenes is claimed to have been the head librarian at the Library of Alexandria and, remarkably, to have calculated the circumference of the Earth.

As far as is known, the belief that the Earth is a sphere has predominated throughout history—after all, it explains why departing ships slip below the horizon and why the Earth casts a circular shadow on the Moon during a lunar eclipse. In his now-lost book *On the Measure of the Earth*, Eratosthenes explained how a particular natural phenomenon started his enquiries. He had heard that the city of Syene (now Aswan) was so far south that at midsummer the sun was directly overhead at noon and an object cast no shadow (in truth the shadow would fall directly beneath it). He also knew this was not the case in Alexandria, where at the same instant an object would have a shadow—a vertical pole would cast one at approximately 7.2 degrees.

Eratosthenes realized that if the Earth is indeed a sphere, this difference in angles occurs because the ground at Aswan "directly faces" the sun at noon at midsummer, whereas at Alexandria the ground is tilted away from direct sunlight due to the curvature of the Earth. Because 7.2 degrees is approximately one-fiftieth of a full 360-degree circle, he concluded that the circumference of the Earth must be fifty times the distance between the two cities. At that time, this measurement was regularly surveyed, so by multiplying the distance by fifty, Eratosthenes was able to calculate a figure for the Earth's circumference which was probably only 5–15 percent in error.

So, it is because of the Greeks that we now think of the Earth as a sphere, a *measurable* sphere, floating free in space—the essential foreknowledge of modern geological science. After Eratosthenes the stage was tantalizingly set for further progress in "Earth-reason," although, as we will see, other ways of thinking were to frustrate that process for many centuries.

At this early stage I should make a fundamental admission: I am not a geologist. I am, in fact, a biologist obsessed from childhood with animal structure, and the fossil history of that structure. For some reason the imprint of an ancient creature emerging from its rocky tomb always caught my imagination, but I soon realized that the creature and the tomb go together. They give each other context and the meaning of each is diminished without the other. Indeed, in the early years, geology and paleontology were really the same discipline—part and counterpart of our early attempts to free ourselves from the Western fallacies of the biblical creation and the Flood. Geologists used fossils to estimate the relative ages of rocks and used rocks to work out the relative ages of fossils. Fascination with the history of life is what eventually, circuitously, led me to write this book.

Throughout my life I seem to have made geological pilgrimages—to the Canyonlands in the U.S., Ararat in Turkey, Italy's Herculaneum, Indonesia's Gunung Agung, Meteora in Greece, and the Red Centre of Australia. Although not my stated aim, I often end up standing on some famous rocky destination. And a few years ago I moved to the middle of Wales, somewhere it is impossible to escape from geology. Near my home a short angry river cuts its V-shaped valley through the Ordovician strata, and some miles further on disgorges into Britain's largest river, the Severn, which then makes a gentler, if sinuous, descent to the sea. Our house is built from different stuff to the houses several miles to the west, and the reason for all these things is geology.

Ironically, the structure of this book on the history of geology reflects the fact that I am neither a historian nor a geologist. It does not follow a neat linear chronology of geological thought from start to finish. Also, it is not split into sections according to the different sub-disciplines of geology, partly because there is disagreement about what those sub-disciplines are. Instead, I have divided it into five sections, based on five geological themes—five ways in which the Earth is far beyond the human scale, five ways which daunted our ancestors but are now, finally, amenable to scientific investigation and human comprehension.

The first of these themes is *time*, and follows the intellectual journey from when we thought the Earth was only thousands of years old, to the present when we believe it to be an unimaginable million times older than that. The second theme is *energy*, the vast capacity of the planet to drive its enormous processes for so long. The third theme is *process*, the ways in which components of the Earth, ranging from the microscopic to the huge, interact to create the forms and phenomena we see around us. Most real geologists would probably prefer me to stop there, but my

The author at White Tank Arch, California, 2015.

geology has two more dimensions to explore. The theme of the fourth chapter is *use*—the ways humans have exploited Earth's constituents and power to create technological civilization, deploying resources that once seemed boundless, but do not seem quite so boundless anymore. Finally, the fifth theme is *life*, tracing the story of geology's inseparable sister discipline: the study of the origins and evolution of life on Earth. So: time, energy, process, use, life.

David Bainbridge, Cambridge 2025

1

TIME

As old as the hills. We take it for granted that "geology" is almost synonymous with "time." If one talks to a geologist about an event one million years ago, they will often consider using the word "recent." Yet for most of the last two millennia this was not the case, and the sense of what we would now call "deep time," and our ability to measure it, have only developed within the last two centuries.

Antelope Canyon, Arizona, 2015.

Testament to the erosive power of water over long periods of time, Antelope Canyon is a deep "slot canyon" in the Colorado Plateau of the southwestern United States.

"The purpose of this dissertation is to provide some estimates of the dating of the Earth, from its origin when it became a world able to sustain plants and animals, and reflect on the changes it has suffered."

James Hutton, 1785

The ancient Greeks' attempts to calibrate the chronology of the Earth now seem very modern. The geographer and historian Herodotus (c. 484–425 BCE) realized that some slow geological processes are amenable to calculation—in particular he realized that the extensive Nile Delta had been formed by the river's slow deposition of tiny particles, and suggested that if the rate of deposition could be quantified, then the age of the delta could be determined. So classical thinkers contemplated an antiquity of Earth beyond everyday human experience, and some even suggested the world is infinitely old.

However, even before the ancient Greeks considered the Earth's origins, the seeds of confusion had already been sown. Legends of an ancient devastating flood have existed almost as long as humans have been writing, perhaps inspired by real catastrophic events in the Black Sea, Mesopotamia, or the Persian Gulf. Co-opted into the Christian Bible, as well as the Torah and, indirectly, the Qur'an, the Flood was a tenacious story, which became

Leonardo da Vinci (1452–1519), *A Map of the Rivers and Mountains of Central Italy, c. 1502–04.*

In this remarkably modern-looking relief map, Leonardo traced the drainage patterns of northern Italian rivers against the local topography.

Fig. 23.

Reculver Church, in 1834.

the basic assumption underlying geological thinking in the Western world well into the nineteenth century. The rest of the Genesis narrative was no less influential, and in the seventeenth century the Archbishop of Armagh, James Ussher (1581–1656), used it to calculate that the World was created in 4004 BCE. Ussher was by no means the first person to make such an assertion and for some time the Earth had been assumed to be too young for geological processes to have occurred to any great extent.

Yet as early as the high renaissance, Leonardo da Vinci (1453–1519) was toying with ideas of deep time. He had observed fossil shells high in the hills, and argued that because mollusks could not outrun a flood, sea levels must once have been higher, a phenomenon which would surely have needed a lot of time to occur. He even proposed the idea that the Earth undergoes cycles of change over time, something which has echoes in aspects of modern geological thinking. And the English natural philosopher Robert Hooke (1635–1703) argued that fossils were the remains of ancient organisms, even if he tried to cram his ideas into a biblical timeframe.

Working mainly in Tuscany, the Dane Niels Stensen (1638–1686), generally known as Steno, made seminal contributions to many areas of geology, but perhaps his most important was to question accepted wisdom concerning what we would now call sedimentary rocks. For example, although he adhered to the Bible's chronology, he argued that the presence of debris and fossils in layers of rock showed that those layers did not all date from the moment of Creation. Instead, he claimed that new layers of rock are deposited on top of older layers—the two do not mix, and the lower layer must already be a solid to support the upper. He posited that although these rock layers may later become confusingly tilted or damaged, the general trend that new layers are laid upon old lies at the heart of Earth's processes.

It could reasonably be argued that the most important person in the history of geology was the Scottish Enlightenment figure James Hutton (1726–1797). A naturalist, physician, and landowner, Hutton put forward many ideas which form the basis of modern geology. Importantly, he based them on actual evidence—and where that evidence was lacking, he went looking until he found it. Hutton is a key figure in "uniformitarianism," the idea that the face of the Earth is the product of long, long periods over

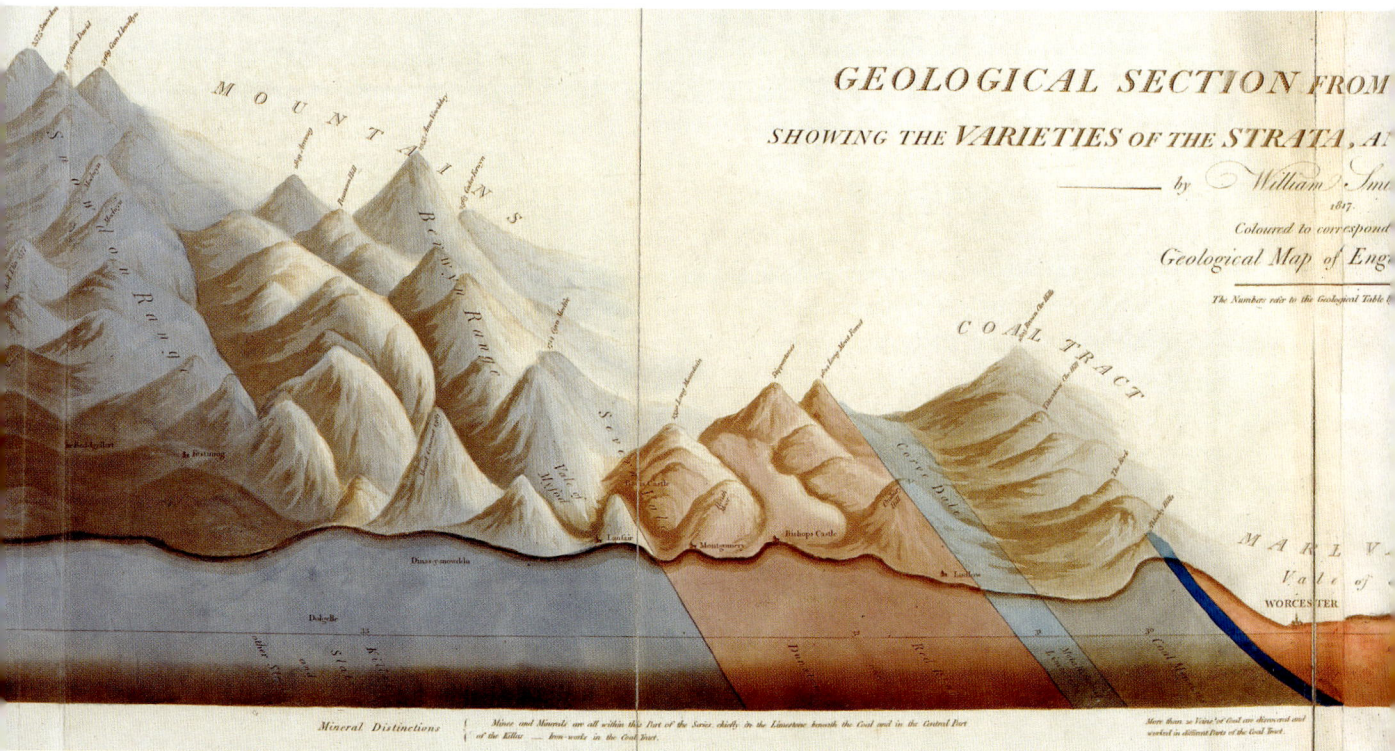

GEOLOGICAL SECTION FROM

SHOWING THE *VARIETIES OF THE STRATA, A*

_____ by William Sm

1817.

Coloured to correspond

Geological Map of Eng

which very slow processes have sculpted its surface. Those processes may themselves remain uniform over time, but they result in continual slow change to the Earth itself.

Key to Hutton's ideas were the "discontinuities" he found in Britain's rocks—places where rock layers had been laid down, presumably by water (he found fossilized ripples), then tilted at an angle, partially worn away, and finally overlain by later horizontal rock layers. To Hutton the time required to form discontinuities seemed incomprehensibly vast, but even so he thought they took just a fraction of the age of the Earth. Indeed, he proposed that the arrangement of land and sea has continually changed over time, with cyclical production of layered sedimentary rocks under water, elevation of those rocks by the Earth's internal heat to form continents and mountains, and the erosion of those highlands to yield the raw materials for the next generation of land-building. Hutton's theories were vast in scope, vast in time, and vastly influential.

The counterbalance to uniformitarianism is "catastrophism"—the idea that Earth's surface was formed mainly by brief violent episodes such as floods, eruptions, and impacts, some of which led to sudden extinctions of many of its inhabitants. An early proponent of this idea was the

William Smith (1769–1839), *Geological Section from London to Snowdon, Showing the Varieties of the Strata and the Correct Altitudes of the Hills,* 1819.

William Smith's survey of the layers of rock ("strata") underlying most of the British Isles was important in establishing the great antiquity of the rocks beneath our feet.

French paleontologist Georges Cuvier (1769–1832), who suggested that the apparent damage and distortion evident in many rocks was evidence of acute, violent processes—although he scrupulously avoided linking these to Bible stories. He also identified what we would now call "gaps in the fossil record," and argued that they indicated short-term change was key to understanding Earth's history. Although catastrophism was to lose ground to uniformitarianism in the nineteenth century, it remains relevant today, as does Cuvier's ominous suggestion that the Earth is subject to cyclical, ever-returning episodes of catastrophe.

During the 1860s, advances in the physical sciences started to contribute significantly to our understanding of the Earth's age. William Thomson, Lord Kelvin (1824–1907), applied prevailing theories of thermodynamics to a model of the cooling Earth. He assumed that the Earth started life as a hot ball of rock, and progressively cooled as heat was conducted to its surface and radiated into space. His calculations produced different results depending on the assumptions made, but it seemed that the planet was at least twenty million years old. However, many uniformitarians worried that even this greatly increased timescale would not be long enough for geological (and evolutionary) processes to have occurred—a concern worsened by very low estimates of the sun's age based on the erroneous assumption that it produced its heat by simple chemical combustion.

The late 1860s also saw the emergence of Dmitri Mendeleev's (1834–1907) first periodic tables of the elements—a classification of the fundamental chemical constituents of the universe based on their properties. Mendeleev's system was an intellectual *tour de force* in which he realized the importance of the repeating patterns of chemical properties of the elements that appear when they are listed in order of mass. He even correctly left gaps in his patterns for elements he suspected had not yet been discovered.

The periodic table remains the central plan of chemistry (see pages 38–39), and it allowed geologists to overcome their previous inability to coherently classify minerals. It was also later extended by the discovery that elements come in variants or "isotopes" of differing mass and that some isotopes—radioactive ones—break down to release energy. Indeed, heat is continually produced within the Earth by these isotopes, and this is one of the reasons why Kelvin's estimates for the age of the Earth were too low.

Research into radioactivity continued apace in the late nineteenth and early twentieth centuries, and the phenomenon had a profound effect on the study of geological time. In 1907 the American radiochemist Bertram Boltwood (1870–1927) used the decay of long-lasting radioactive isotopes (in this case uranium) as a chronometer to measure the age of rocks, and estimated Earth to be up to 2,200 million years old. Later refinements have

Natural Bridges National Monument, Utah, 2015.

Utah's first National Monument, the Natural Bridges are enormous arches formed when small streams slowly undercut existing ridges of rock.

suggested an age of 4,540 million years, so Boltwood's first estimate was a good one. Radiometric dating has proved to be the holy grail of deep time ever since, allowing accurate measurements backed by hard physics.

Once radiometric dating had fixed the *extent* of geological time, later study focused more on is its *nature*. In 1929 it was discovered that the Earth's magnetic field reverses in an episodic, although irregular, fashion every couple of million years or so. In the 1930s the possibility of cycles related to changes in the Earth's rotation came to the fore, and later fed into theories of mass extinction. Indeed catastrophism has made something of a comeback in recent decades as worldwide extinction events have been convincingly linked to cataclysmic volcanism or asteroid impacts. In fact, contemporary geology views occasional catastrophes as part of the ongoing processes which sculpt the Earth—so catastrophism has really become just another part of uniformitarianism.

Time continues to obsess geologists. On pages 52–53, we encounter the ongoing race to find the oldest Earth minerals, as well as the discovery of far older material, forged elsewhere in the Universe. And to complete the picture, we now finally have a detailed, robust chronology of the Earth.

ABOVE Gilgamesh slaying the Bull of Heaven, cylinder seal, c. 2250–1900 BCE.

The narrative of the Flood dominated geological thinking well into the nineteenth century. While the Genesis tale of Noah and his ark is well known, it is almost certainly an adaptation of an earlier story—also included in the *Epic of Gilgamesh*, written in Mesopotamia, perhaps around 1800 BCE. In the *Epic*, supposedly set around 2700 BCE, a character called Utnapishtim is instructed by the god of wisdom to build a boat to particular specifications, that he might ride out the coming inundation. For many centuries in parts of the world dominated by Christianity, geological phenomena were often misinterpreted as having resulted from the Flood, a tendency which did untold damage to the development of geological science. However, some have suggested the Flood story may actually derive from folk memories of real events taking place at the end of the ice ages or soon after—floods of the Black Sea or Persian Gulf, or catastrophic collapses of glacial dams in the wider region.

RIGHT *In principio creavit Deus caelum et terram* (the Creation of Heaven and Earth), from the Aberdeen Bestiary, c. 1150.

"In the beginning, God created the heaven and the earth." The book of Genesis does not state when the Creation took place, although theologians often attempted to calculate the age of the Earth by adding up the intervals between events listed in its narrative. Many came up with estimates of a few, or several, thousand years—a relatively young Earth, aligned to the time course of human history, but apparently too short to allow the achingly slow processes which some thinkers believed underlie geological change. Ironically, in this illustration God stands atop two of four pre-existing rocks, possibly representing his use of the four elements of earth, air, fire, and water to create his world.

Hossein Behzad (1894–1968), *Ibn Sina* **(Avicenna), 1950.**

Working mainly in Iran, the philosopher Ibn Sina (*c.* 980–1037) was a central figure of the Islamic Golden Age, whose fame and influence spread far into Europe— he is name-checked in Geoffrey Chaucer's fourteenth-century *Canterbury Tales.* Distant from the constraints of Christianity, he freely speculated on the nature of geological processes—how stones are made and how they form mountains, for example. In particular he argued that the Earth's surface is the creation of ongoing geological processes which take enormous amounts of time—erosion of mountains by water, and the deposition of sediments in low-lying areas.

Leonardo da Vinci (1452–1519), *Madonna of the Rocks, c.* **1486.**

Leonardo revisited his favorite visual tropes again and again—swirling water, folded drapery, and, in this case, rock formations. However, his interest in geology was not just a casual artistic preference. It is known that Leonardo climbed the hills and mountains of the Italian Alps and visited Tuscany collecting fossils, and he was clear in his opinion that such finds were the remains of ancient animals rather than capricious, spontaneous creations of Earth. He noted, for example, that fossil shells show signs of parasitism. He argued that the presence of fossils of marine mollusks, and possibly even whales, high on Italian hills was a sign not of a biblical deluge, but the elevation of ancient seabeds. He described lines of deposition of such fossils, what we would call "strata," and suggested that they were laid down sequentially over time. He also argued that flowing water (see page 14) was crucial in the achingly slow destruction of mountains.

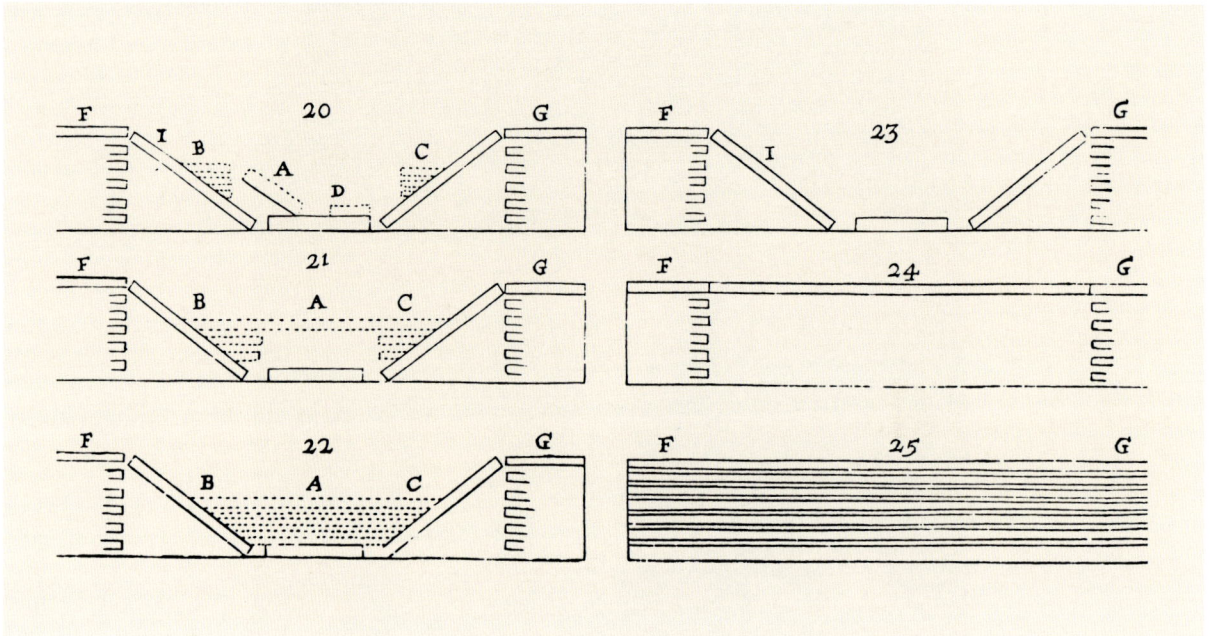

Niels Stensen (1638–1686), the Arno River valley, plate xi from *The Prodromus to a Dissertation Concerning Solids Naturally Contained Within Solids*, 1669.

Born in Copenhagen to a wealthy Lutheran family, Stensen, or Steno, studied initially as an anatomist in Amsterdam—discovering, among other things, the small tube that drains saliva into the mouth. However, after traveling to Italy, his interest turned to fossils, as well as crystals and rocky intrusions which may be found within pre-existing rock layers.

His main contribution to geology was to characterize the time sequence within which rocky phenomena appear. He argued that what we would now call sedimentary rocks are always laid down in horizontal layers, new layers always on top of old, but that later processes can tilt these layers, or inject fossils or other types of rock into them in clearly identifiable ways.

In what was, for the time, an unusually schematic diagram, Steno here explains his ideas concerning how the valley of the Arno—the great river which flows through Florence—has developed over long stretches of time. Confusingly, the images should be read in reverse numerical order. The illustration numbered 25 shows the perfect layering of rocks present after the Creation. In illustration 24, the great flood has washed away underlying layers, undermining the surface, which then collapses down in illustration 23 to create the valley. Thereafter more layers of rock accumulate in this depression in illustration 22, only to be "re-undermined" in illustration 21, giving rise in illustration 20 to the modern-day jumble of layers and ages of rock we see today.

Jean-Étienne Guettard (1715–1786) and Philippe Buache (1700–1773), *Mineralogical Map on the Nature and Situation of the Lands Crossing France and England,* **1746.**

Philippe Buache, officially the geographer to French king Louis XV, and Jean-Étienne Guettard, geologist at the French Académie Royale des Sciences, produced one of the first modern geological maps. Impressive in scope, it traces the chalk-bearing regions around the Paris Basin and extends them across into England. The implication was that rocks within the chalk ring are younger than the chalk, and rocks outside the ring are older. Later, the chalk-bearing, or "Cretaceous" period was named after these deposits.

Johann Gottlob Lehmann (1719–1767), illustration from *Versuch einer Geschichte von Flötz-Gebürgen* (Essay on an History of the Stratified Mountains), 1756.

Lehmann was a German medical doctor and mining engineer who worked initially in Saxony and Prussia, and later in Russia as far east as the Ural Mountains. He realized that mountains that appear superficially similar can have very different internal structures—some are made from layered rocks, whereas others have less internal structure but are often laced with veins of valuable ores. Extending the work of Steno, Lehmann became fascinated by the orderly laying down of layers of sedimentary rock in strict historical sequence—sometimes characterizing rock formations with tens of different, neatly stacked layers.

Giovanni Arduino (1714–1795), stratigraphic section of the Valle dell'Agno near Vicenza, Veneto Region, with north to the left, 1758.

The "Father of Italian Geology," Giovanni Arduino was the first to use the layering of rock to develop a chronology of geological time which resembles the one we use today. He sub-classified what we now know to be the last six hundred million years or so into "Primary," "Secondary," and "Tertiary" eras, as well as using the term "Quaternary" for volcanic rocks. The first three correspond approximately to what we now call the Paleozoic (540–250 million years ago), Mesozoic (250–66 million years ago), and Cenozoic (66 million years to the present) eras. His term Tertiary survived almost until the present day, as did Quaternary, although with a different meaning. It now refers instead to the relatively short period after the Tertiary. This sketch illustrates Arduino's concept of how the landscape of Tuscany is formed by multiple overlapping layers of sedimentary rock.

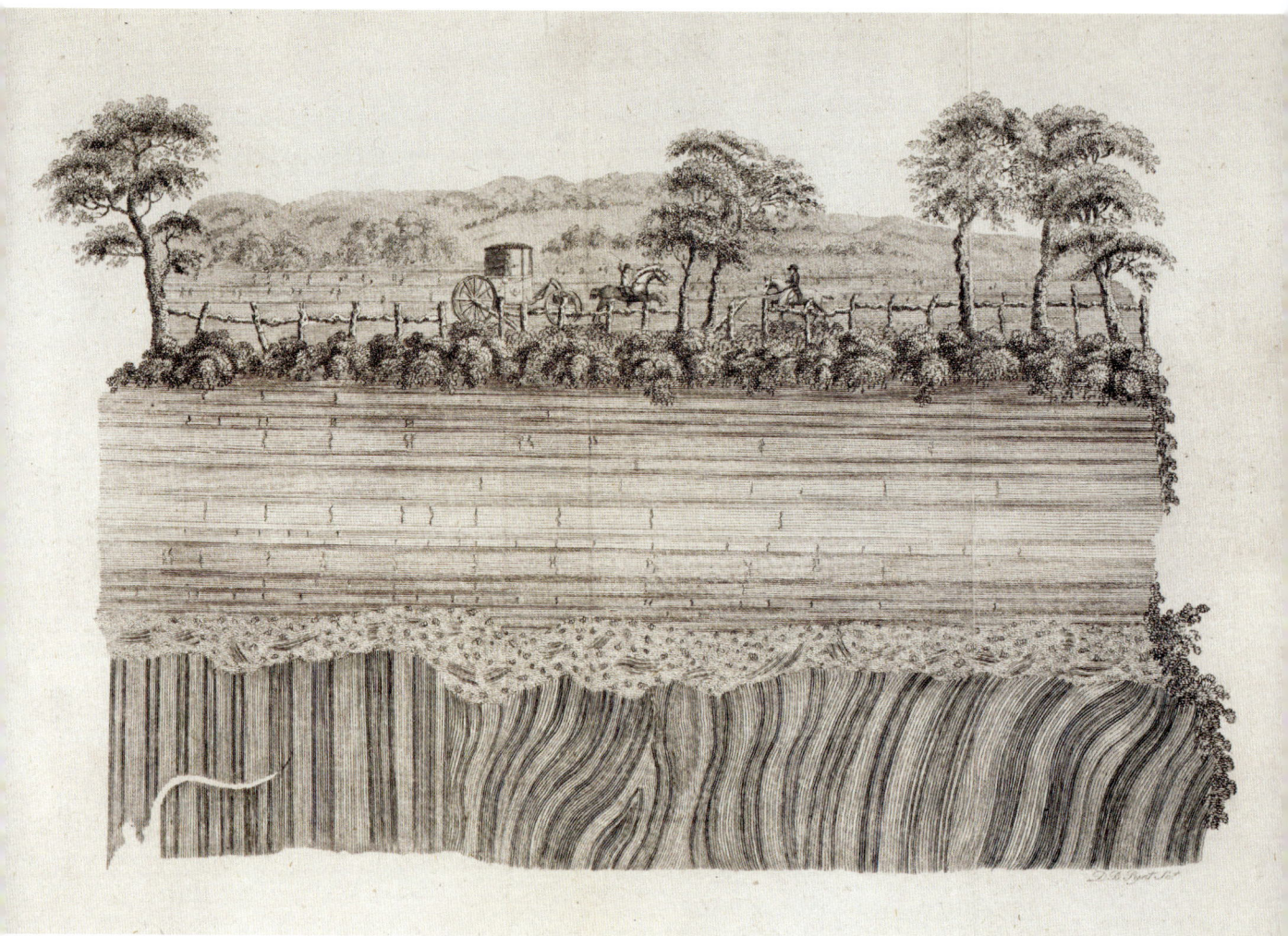

John Clerk of Eldin (1728–1812), "Unconformity at Jedburgh," 1787, in later editions of James Hutton (1726–1797), *Theory of the Earth, 1788.*

Although it was by no means his only contribution to geology, our modern conception of deep time derives almost entirely from the work of the Scottish polymath James Hutton (see page 16). In particular, he drew on the evidence of "unconformities" to support his theory that Earth is unimaginably old. An unconformity occurs where layers of sediment have been laid down, presumably under water, raised above those waters, partly eroded away, tilted so that they lie aslant, settled downward underwater once more, covered by new horizontal layers of sediment, and finally are raised again into view—a sequence of events which surely requires enormous expanses of time.

Siccar Point, Scotland, 2024.

Although not the first discontinuity Hutton discovered, Siccar Point on the coast of Berwickshire, Scotland, is perhaps the most important—indeed possibly the most significant place in the history of geology. Here oblique bands of grey shale are overlain by red sandstone in a way that is, as Hutton wrote, "washed bare by the sea," and which in a companion's words was like "looking into the abyss of time." Hutton claimed that repeated episodes of collapse, uplift, displacement, and distortion were driven by the heat deep in the Earth. In his wonderfully titled *Theory of the Earth; or, an Investigation of the Laws Observable in the Composition, Dissolution, and Restoration of Land upon the Globe*, Hutton argued that ours is an ancient planet in continual flux. Almost suggesting that geological time is infinite, he wrote, "The result, therefore, of this physical enquiry, is that we find no vestige of a beginning, no prospect of an end."

William Smith (1769–1839), *A Delineation of the Strata of England and Wales, with Part of Scotland*, 1815.

William "Strata" Smith was a surveyor and canal engineer whose work sent him all over England and Wales. Digging canals requires an excellent understanding of the layers of rock, or "strata," in which one is working, but Smith's inquisitiveness took him further than just the practical knowledge he needed. He realized that across the island coal seams are consistently flanked by the same sequences of strata, above and below. Although he did not extrapolate his findings further afield, nor break from the tyranny of biblical timescales, he cemented the idea that the rocks represent a consistently ordered catalog of the constituents of the Earth. He was also instrumental in linking the study of rocks to the fossils they bear (see pages 216–217).

His greatest legacy is this vast and beautiful map—original copies measure more than 1.8 × 2.5 m (6 × 8 ft). His use of color and implied relief far surpass anything that went before, and cemented the idea that the British are unusually blessed by the geological variety of their small island. Unfortunately, producing these large maps proved to be part of Smith's financial undoing. Along with moving from full-time to freelance work, and some poor investments, the cost of creating these geological artworks was to lead to a stint in a debtors' prison and an impecunious old age.

Great Britain in the early nineteenth century was indeed an ideal place and time to catalog the Earth's crust—the industrial revolution was driving a practical need to understand the rocks through which engineers excavated, tunneled, and mined. Also, fortuitously, most of the last half-billion years of geological history is conveniently exposed across the island in diagonal stripes: southwest-to-northeast bands stacked in series from the ancient Scottish Isles to the recent Thames Valley, in England.

Diving into deep time

CHARLES LYELL (1797–1875)

The geologist Charles Lyell was a towering and influential figure in Victorian science. Born in Scotland, his family moved to the south coast of England where he grew up and developed a keen interest in the animal and mineral world of the New Forest. He studied law at Oxford University but soon became fascinated by lectures on the nascent science of geology, and attempts to reconcile it with biblical timescales.

Lyell continued his studies and received his license to practice law. However, with considerable funding from his father, he was soon distracted by the opportunities to study geology around London—indeed he was later to travel to many of the classic destinations on any geological grand tour: Sussex and Norfolk in England, Italy, the Pyrenees in France and Spain, and Switzerland, as well as much of the eastern United States and Canada.

Italy's Bay of Naples was particularly important in the development of Lyell's ideas. On the north side of the bay are the Phlegraean Fields, or Campi Flegrei (see also page 63), a caldera, or low, shallow bowl-shaped volcano. Lyell became fascinated by the evidence of its cyclical rising and falling. In particular, there are the three remaining columns of the Roman-era Macellum of Pozzuoli, which now stand proud above the surrounding landscape, yet bear indubitable

Mount Etna, Sicily, 2008.

evidence of boreholes created by marine bivalve mollusks. Lyell realized that, since their erection two thousand years ago, the land beneath these columns must have slumped down into the sea for a period, and then been re-elevated to its current position—a cycle of geological activity lasting only two thousand years.

Indeed, Lyell was fascinated by timescales—for example, he traced the evidence of British churches being swamped by soil or lost to the sea over periods not much longer than the human lifespan. Conversely, he estimated the age of Mount Etna in Sicily to be approximately one hundred thousand years, by extrapolation from the rates of deposition of ash and lava. In this way, he amassed copious evidence of geological processes indicating that the Earth is extremely old and subject to processes which do not change fundamentally over time—he was very much Hutton's uniformitarian scientific heir.

Lyell's influence was also to prove crucial to the most important discovery in biology. Charles Darwin (1809–1882) wrote at length about how Lyell's concept of an ancient Earth influenced his own thoughts on how animal and plant species change over time—it could be claimed that Lyell's chronology gave evolution sufficient time to occur. Lyell also played a key part in the announcement of the theory of natural selection itself to the world. He was aware that Alfred Russel Wallace (1823–1913), working far away in the Dutch East Indies, and Darwin had arrived at essentially the same mechanistic explanation for how evolution takes place, and it was he who mediated between the two so they could co-publish a single historic paper on their new theory.

Charles Lyell, "Macellum of Pozzuoli," frontispiece to *Principles of Geology*, 1830.

"Lyell realized that, since their erection two thousand years ago, the land beneath these columns must have slumped down into the sea for a period."

Spirifer Hawkinsi, Morris + Sharpe, + *Orthis Sulivani,* M+S.
Lower Devonian
Fig.ᵈ Morris + Sharpe: Quart. Journ. Geol. Soc. 1846, Vol. 2 pl.11, f.1a.b., pl.10 f.1c, pp.275.276.,
[B. 17794] ~~13.CC.~~ + Charles Darwin Collection
Falkland Islands
Transᶜ M.P.G. 1880.

A specimen containing multiple species of Lower Devonian brachiopods collected by Charles Darwin (1809–1882) at Port Louis, East Falkland, 1833.

Many of Charles Darwin's contributions to science were rooted in observations made during the famous voyage of HMS *Beagle* between 1831 and 1836. He carried a copy of Lyell's *Principles of Geology* with him on the trip and was fascinated by the idea that land could rise and fall over time. While in South America, he observed the fossil remains of entire oyster beds hundreds of meters above the sea, and scattered shells more than 3,600 m (12,000 ft) up. This indirect evidence of "uplift" was bolstered by a visit to Concepción in Chile, soon after its devastating 1835 earthquake—it was clear that the seabed had been lifted many feet during the quake so that mussel shells could be found rotting in the sun, high and dry above sea level. Darwin was later to use ideas of uplift and subsidence to explain the formation of coral atolls (see page 219).

Edmund Mojsisovics (1839–1907), *Die Dolomit-Riffe von Südtirol und Venetien: Beiträge zur Bildungsgeschichte der Alpen* (The Dolomite Reefs of the Südtirol and the Veneto: Their Contributions to the Formation of the Alps), 1879.

The German explorer Ferdinand von Richthofen (1833–1905) was the first to realize that there is a vast and dramatic display of seabed uplift in the heart of Europe— the Dolomite Mountains in northern Italy. He mapped the valley of Predazzo and found limestone boulders and huge peaks made of other carbonate rocks, exposed in sloping cross-sections across entire mountainsides. He argued that this mineral composition indicated that the Dolomites were built from ancient reefs and atolls, thrust high into the air by unimaginable forces.

ABOVE The Grand Canyon viewed from its north rim, Arizona, 2008.

LEFT John Wesley Powell (1834–1902), "Section of a Wall in the Grand Canyon," from *Exploration of the Colorado*, 1875.

The story of John Wesley Powell's descent of the Colorado River in small wooden rowing boats is one of remarkable courage and fortitude. Despite the fact that he had lost an arm during the American Civil War at the Battle of Shiloh, and that the proposed journey was through largely unexplored territory, in 1869 he set out with ten men and four boats from Green River in Wyoming with the aim of riding the river to its confluence with the Grand River (the old name for the Colorado). Two years later he continued the dangerous journey and despite arguments, desertions, and disasters, his party survived a traverse of the river's remote and dramatic Utah reaches, and through the Grand Canyon in Arizona. Despite his desperate struggle for survival, Powell recorded many aspects of the area's geology. The Grand Canyon was cut through the Colorado Plateau as it was uplifted over a period of perhaps 60 million years—the river cutting ever deeper until the canyon is now 1,800 m (6,000 ft) deep. The flanks of this great incision in the Earth now reveal sediments laid down (and tilted, and eroded) over a period of 2,000 million years—nearly half the age of the planet.

Elements, octaves, and time

THE GENIUS OF MENDELEEV

Throughout history humans have pondered the nature of the basic building blocks of matter. Very few of what we now call chemical elements were known in their pure form to the ancients—probably only the yellow sulfur which condenses at volcanic vents, and a few of the more readily purified metals. Certainly, the science of geology could not progress far without a better understanding of the chemical basis of the minerals beneath our feet.

Toward the end of the eighteenth century, chemists had considerable evidence that everyday matter is formed of vast numbers of tiny atoms, and that these atoms come in many varieties, or elements. Indeed, by the start of the nineteenth century the weight of each type of atom could be measured, and this led the English chemist John Dalton (1766–1844) to suggest that the elements could be usefully listed in ascending order of weight. He also proposed that atoms of particular elements would combine only in strict ratios—for example, one volume of oxygen gas reacts with two volumes of hydrogen to create water, H_2O.

As more elements were discovered, scientists started to discern patterns in the ranked lists of chemical elements. In 1865, John Newlands (1837–1898) formulated

LEFT John Dalton (1766–1804), table of the elements, 1808.

RIGHT The periodic table.

The lighter elements are at the top of the table and the heavier ones at the bottom. In this version the elements are numbered by their "atomic number," the number of positively charged protons in their nucleus.

his "Law of Octaves," which states that the properties of elements repeat every eight elements in the list. In 1869, the Russian, Dmitri Mendeleev went further and used this apparent cyclicity (periodicity) of the properties of elements to create the first recognizable periodic tables.

No other science can be represented as succinctly as the periodic table summarizes chemistry—the figure below depicts the first ninety-two elements of the table. Columns contain elements of similar properties—for example shiny copper, silver, and gold (Cu, Ag, and Au; elements 29, 47 and 79). The region from helium to calcium (He to Ca; elements 2–20) is where elements' properties repeat every eight elements—Newlands' octaves—whereas the cycles become longer lower in the table.

Mendeleevs's periodic table explains chemistry, and it sparked a revolution in geology. Finally, the constituents of minerals could be determined and understood, and a coherent hierarchy of geological entities could be established, from tiny atoms, through crystals and minerals to large rocks. Moreover, this chemical revolution led to great progress in our understanding of geological time. When did the elements form? We now know that hydrogen and helium were formed in the early universe and still make up 98 percent of the mass of its matter. When and why did particular elements come to form the Earth? In fact, oxygen, aluminum, and silicon turn out to be the most common elements in the planet's crust, but were produced within long-dead stars whose carcasses were recycled to form the solar system. And as we will discover (see page 45), the unstable properties of some elements' atoms would soon become the primary way we measure geological time.

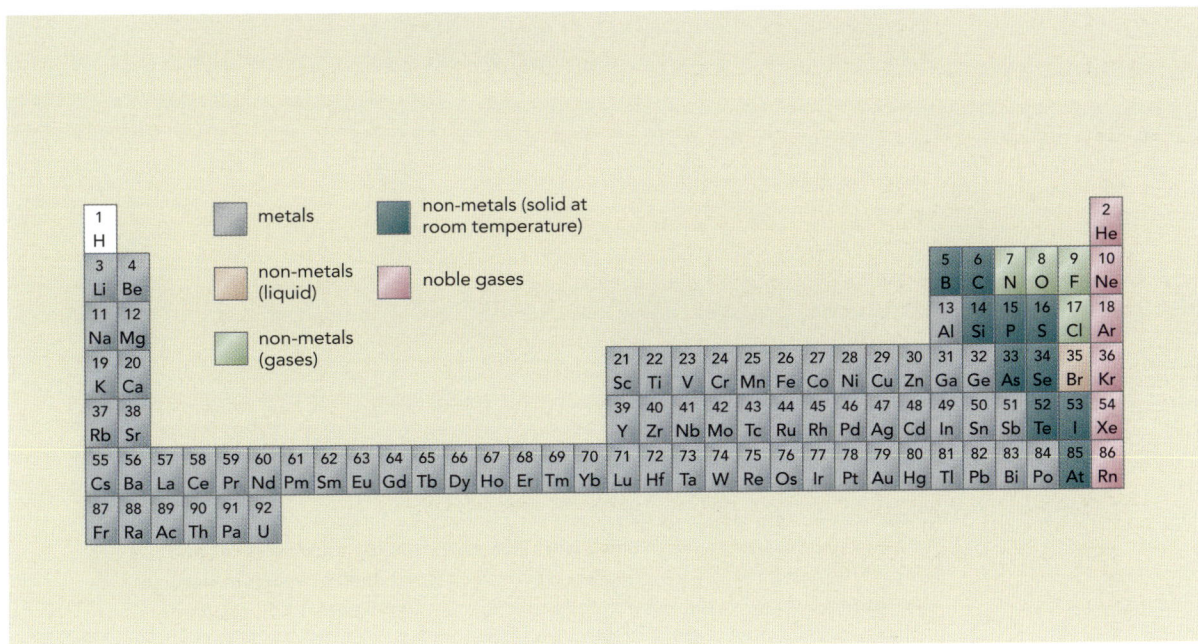

William Thomson, Lord Kelvin (1824–1907), demonstrates to students in Glasgow University. Artist and date unknown.

By the middle of the eighteenth century, physicists were applying recent advances in thermodynamics to calculate the age of the Earth—often by assuming that it formed as a hot, molten ball, and since has progressively cooled to its current temperate state. The best known of these attempts, although not the first, was by the British mathematician and physicist Lord Kelvin. Making various assumptions —the initial temperature of the globe, the rate at which heat is conducted outward from its center and lost at its surface, along with measurements of temperatures at different depths under the ground—his initial estimate was twenty to forty million years. This was far too long for those who believed in the historicity of the Bible, but perhaps too short for evolution to have produced the diversity of life on Earth. We now know this estimate to be roughly one hundred times too low—partly because Kelvin did not know that the Earth's heat is continually replenished by the decay of radioactive atoms within it, but possibly more importantly because he did not realize that much of the heat flow in the Earth is due not to conduction but to convection—bulk circulation of its viscous interior.

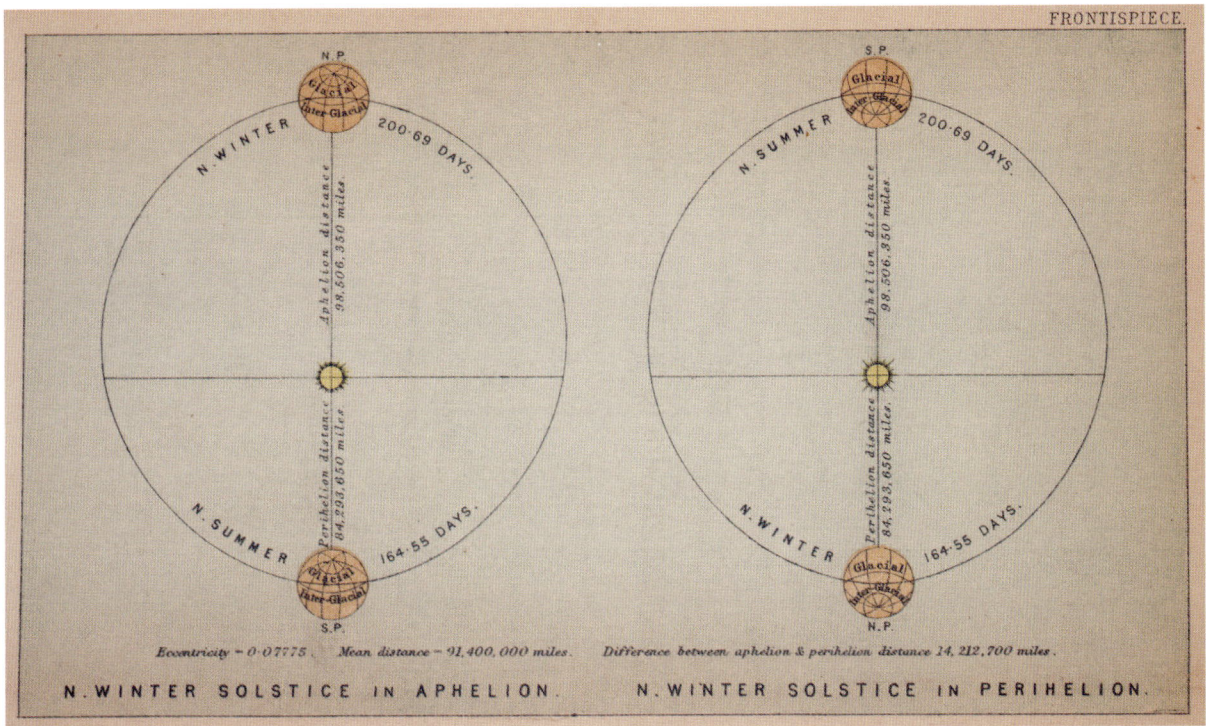

N·WINTER SOLSTICE IN APHELION.   N·WINTER SOLSTICE IN PERIHELION.

James Croll (1821–1890), "Earth's Orbit When Eccentricity is at its Superior Limit," frontispiece to *Climate and Time, in their Geological Relations*, 1875.

James Croll came from a poor background in Scotland and struggled financially all his life—often working as a casual laborer. However, he developed a fascination with science as a child, and in particular with the possibly cyclical nature of the Earth's processes. When he was appointed maintenance supervisor of Andersonian College in Glasgow, he gained access to its library, and he started to focus on one of the major questions at the intellectual intersection between geology and climatology—why periods of glaciation have come and gone during the planet's history.

He realized that one possible source of cyclicity over time is changes in the Earth's orbit around the sun. The orbit is not a perfect circle, but is stretched into an ellipse—the orbits are slightly slimmer in the left-to-right direction in the image, and the Earth is further from the sun at the "top" of the ellipses than at the bottom. Croll studied how the stretch of that ellipse has changed over time and whether the earth's own axis of rotation has slowly wandered. Croll corresponded widely, including with Charles Lyell and Charles Darwin, but his ideas were not generally accepted during his lifetime—partly because they produced too early a date for the end of the last ice age. However, corrections to his estimates were soon made and it is now clear that Croll instigated the study of how celestial cycles affect the Earth.

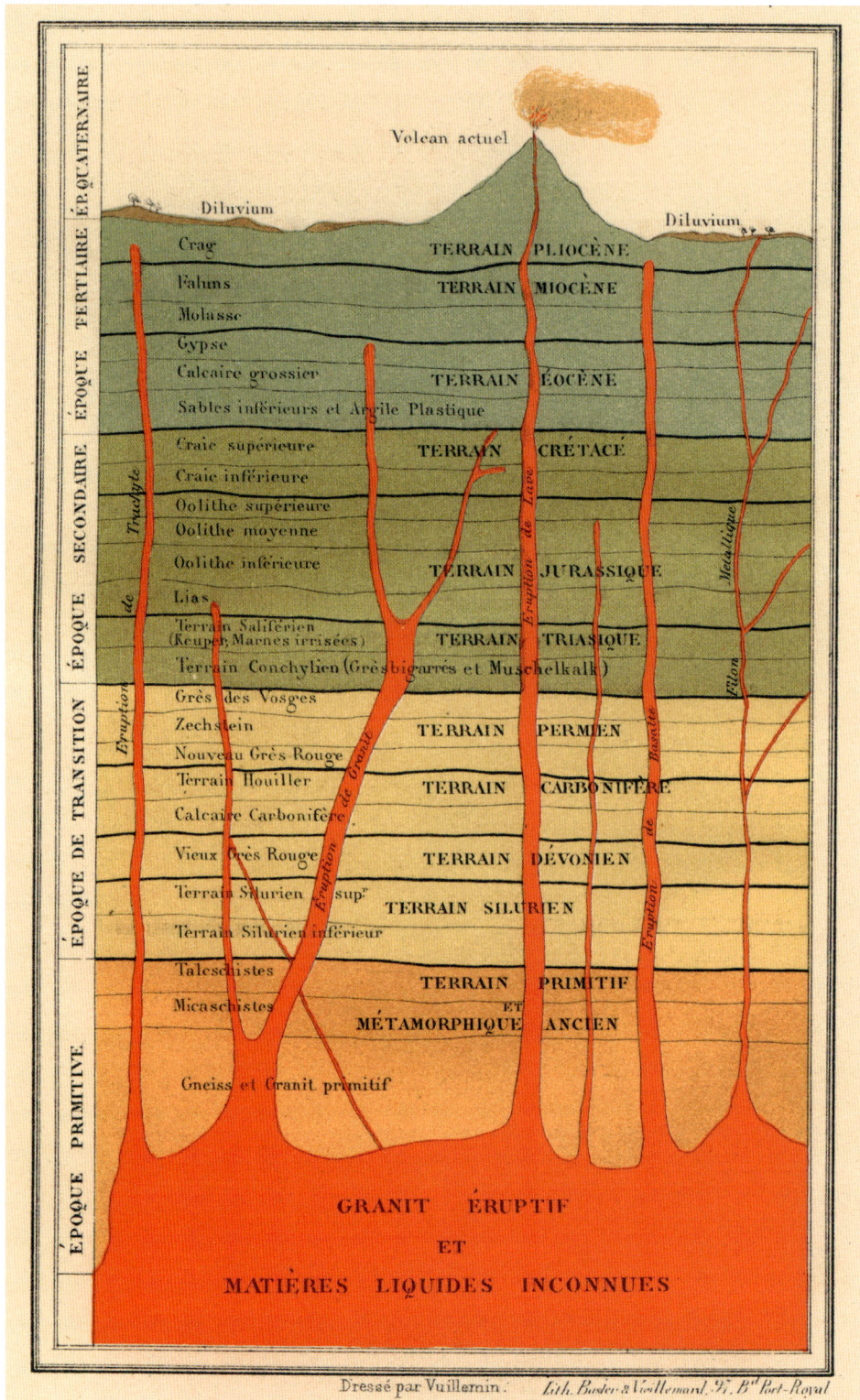

LEFT Louis Figuier (1819–1894), illustration showing eruptive granite and unknown liquid materials, frontispiece to *La Terre Avant Le Deluge* (*The World Before the Deluge*), 1871.

ABOVE Levi Walter Yaggy (1842–1912), "Geological Chart," from *Yaggy's Geographical Portfolio*, 1893.

While competing theories of the Earth were still coalescing, the nineteenth-century obsession with orderly classification led to the naming of the geological periods we still use today, albeit in refined form. Britain is blessed with regular "stripes" of rock bands conveniently arranged from ancient northwest to recent southeast, and in the early nineteenth century it had been the location of both an industrial revolution and a scientific enlightenment, so it is little surprise that the words Cambrian, Silurian, Devonian, Carboniferous, and Ordovician relate to regions or ancient tribes of Britain (all except the last were coined in the 1820s and 1830s; Ordovician appeared later to solve an inconsistency created by earlier feuds between British geologists). The French also had some influence, with Jurassic and Cretaceous, while Triassic relates to Germany and Permian to Russia.

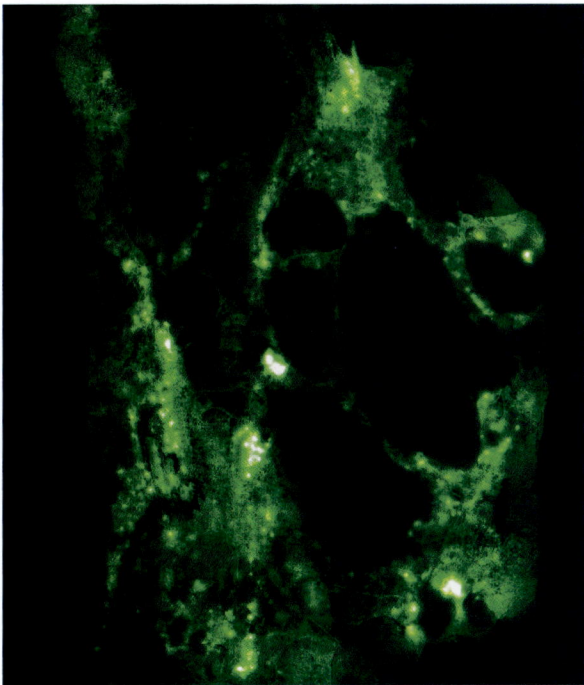

LEFT A sample of uranium ore conglomerate from Ontario, Canada. Date unknown.

ABOVE Radiation from the same sample of uranium ore conglomerate, captured on photographic film.

In the early twentieth century, major advances were made in our understanding of radioactivity. It was being recognized that each chemical element can exist in multiple forms—with near-identical chemical properties. However, for each element some of these forms are relatively stable, whereas others are prone to spontaneous radioactive decay, unpredictably ejecting a spray of particles and often turning into a different element in the process.

We now know these different variants of an element as "isotopes," and that they vary not in the number of positively charged protons that defines each element (see page 39), but in the number of uncharged neutrons in their nucleus. For example, the ninety-second element, uranium, exists in several forms. Uranium-238, which contains 92 protons and 146 neutrons, is relatively stable—only about half of a sample existing since the origin of the Earth would have decayed away by now. In contrast, half of a sample of uranium-232 (92 protons and 140 neutrons) would be gone after seventy years, that isotope's *half-life*. The images show a fragment of uranium-bearing rock, and the ghostly image left on a photographic plate by the radioactive emissions of that uranium.

Although the decay of individual atoms cannot be predicted, the amount of time for half of a sample of an isotope to decay is consistent. One half of uranium-232 is left after seventy years, one quarter is left after one hundred and forty years, one eighth after two hundred and ten years, and so on.

The American radiochemist Bertram Boltwood (1870–1927) was one of the first to realize that this predictable decay of nuclear isotopes might be used to determine the age of rocks. He worked on uranium, radium, and their decay products in a variety of rocks and in 1907 published his estimated age of the Earth—2,200 million years. This estimate seemed excessively high to many scientists at the time, although we now know it to be roughly half the planet's actual age.

However, Boltwood was a radiochemist with little interest in geology, and it fell to Arthur Holmes (1890–1965), working at Imperial College London, to finesse these techniques and realize the full impact of radiometric dating. Holmes was gripped by the fact that at last an accurate clock of the Earth might be within his grasp, and seemed positively excited by the possibility that the Earth is far older than tens of millions of years. Indeed, the old Cambrian period was now seen to have been preceded by the far longer and tantalizingly mysterious Precambrian—a vast uncharted expanse of time that geologists are still struggling to get to grips with.

Radiometric dating revolutionized the measurement of geological time, and remains *the* key tool for calibrating and conducting geological time measurements.

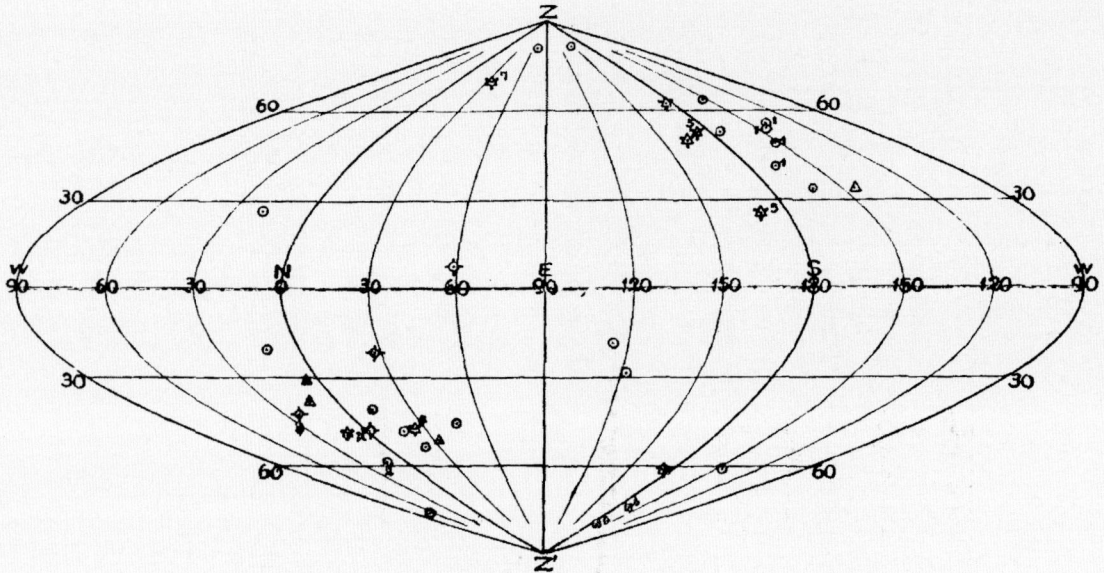

Monotori Matuyama (1884–1958), diagram showing the direction of the north magnetic pole of basalt, from "On the direction of Magnetization of Basalt in Japan, Tyôsen and Manchuria," *Proceedings of the Imperial Academy of Japan*, vol. 5, pp. 203–5, 1929.

Born in Oita in southern Japan, Monotori Matuyama (1884–1958) made his greatest contribution to geology while working at Kyoto Imperial University. For some time there had been a suspicion that the polarity of the Earth's magnetic field reverses from time to time, but Matuyama was the first to conduct a systematic study of the magnetization of rock strata over a wide area—in this case Japan, Korea and Manchuria.

He discovered that the planet's north and south magnetic poles have indeed swapped repeatedly over geological time. This map projection of the Earth's entire surface shows the position of its north magnetic pole at different times in the past, each calculated from a different rock sample. The north magnetic pole can be seen to have been located in one of two diametrically opposite regions—one near the position of the current magnetic north magnetic pole, and the other region near current magnetic south. On average, the magnetic field of the planet reverses every half-million years or so, although reversals occur very erratically. We now believe that each magnetic reversal takes place over a period of only thousands of years.

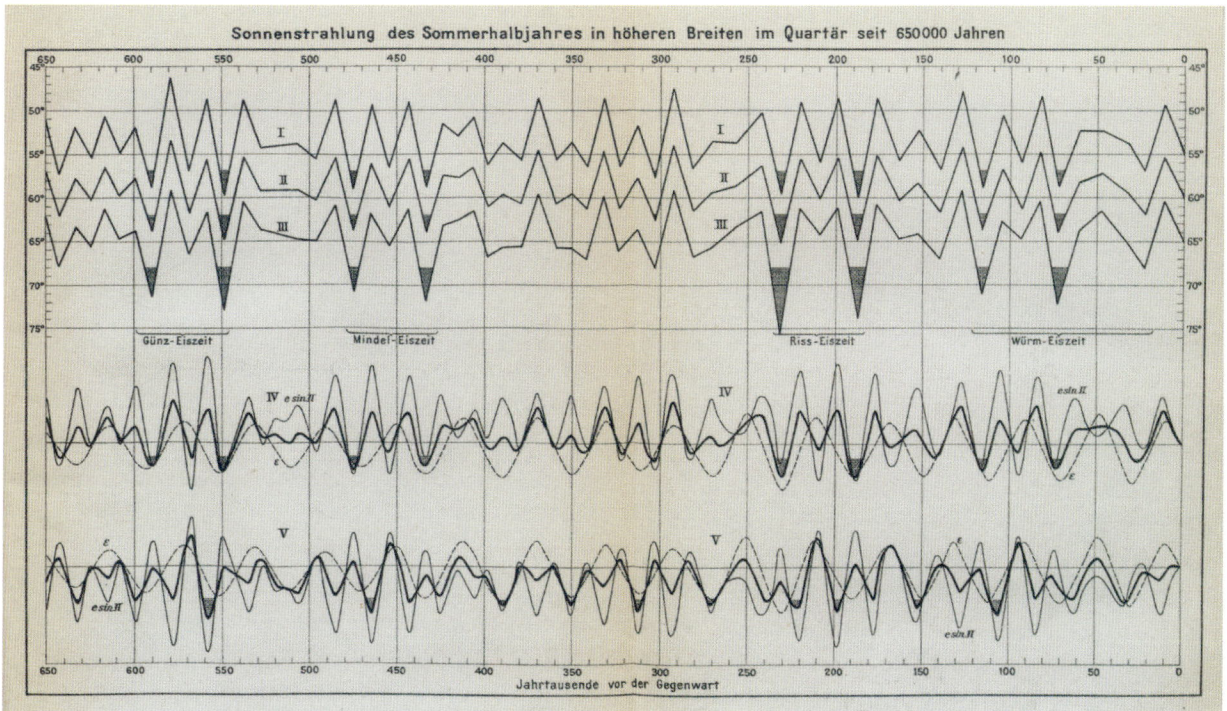

Sonnenstrahlung des Sommerhalbjahres in höheren Breiten im Quartär seit 650000 Jahren

Wladimir Köppen (1846–1940) and Alfred Wegener (1880–1930), "Solar Radiation During Summer Half-Years in Higher Latitudes in the Quaternary Over the Last 650,000 Years," from *Die Klimate der geologischen Vorzeit* (The Climates of the Geological Past), 1924.

The Serbian mathematician, Milutin Milankovitch (1879–1957) made a pivotal contribution to our understanding of how Earth's orbit and rotation can change its climate and levels of glaciation, and thus affect its geology in a cyclical manner. Milankovitch studied the work of Croll (see page 41) and greatly refined it. Using improved data he established how changes in the ovalness of the Earth's orbit, tilting of the angle of the planet's axis of rotation, and drift of the axis akin to the wobble of a spinning top (eccentricity, obliquity, and precession, respectively) can change the intensity of sunlight reaching the Earth's surface—especially in the mid-latitude continents

which dominate the northern hemisphere. These changes are now called "Milankovitch cycles" in his honor, and they can be used, as he suggested, to extrapolate both backward and forward in time to predict glaciations.

The above image is by two of Milankovitch's supporters, both now famous for other reasons: Köppen for giving his name to today's most commonly used system of climate classification, and Wegener for championing the theory of continental drift (page 81). The upper traces represent the amount of summer sunshine at three northern latitudes (I, II, III) across the last 650,000 years. These traces dip "into the grey" during four distinct ice ages—*Eiszeit* in German. The lower two traces represent the complex rotational and orbital cycles of the Earth's motion over the same time period, suggesting that these cycles drive the changes in solar radiation, and thus the ice ages themselves.

NASA, counting craters on the Moon, 2010.

During the middle of the twentieth century, it was realized that the Moon could tell us a great deal about the geological history of the Earth, and perhaps even its origins. It had been known for some time that our moon is atypical—it seems, for example, strangely large to be circling a planet as small as the Earth. In 1946, the Canadian-American geologist Reginald Aldworth Daly (1871–1957) suggested that our oversized moon was formed when a large object smashed into the proto-Earth and ejected it into orbit.

Although this theory sounds outlandish, it is now the most popular explanation for the formation of both the Earth and the Moon. Indeed there is considerable evidence to support it: later return of moon rocks has shown that the two bodies have similar constituents, the Moon's rocks seem to have formed some considerable time after the solar system formed, the moon has an unexpectedly small dense core, and lacks chemicals that would have boiled away in a high-energy impact. The collision hypothesis is also supported by computer simulations, and also explains some unusual features of the two bodies' rotations and orbits. We now think the impactor was around the size of Mars, and that most of it was incorporated into our own planet. Importantly, it is thought that the presence of such a large satellite has stabilized the Earth's rotation, making it more conducive to the evolution of life.

The Moon and its craters are also claimed to offer a useful method for measuring geological time. In the "crater counting" method, a freshly created surface on a planet or satellite is assumed to accumulate crater impacts at a predictable rate, allowing its age to be estimated from its density of craters. The Estonian astronomer Ernst Öpik (1893–1985) was fascinated by collisions between celestial bodies and, in 1971, was the first to use crater counting to calculate a geological age, estimating the age of the Moon's Mare Imbrium to be around 4,500 million years.

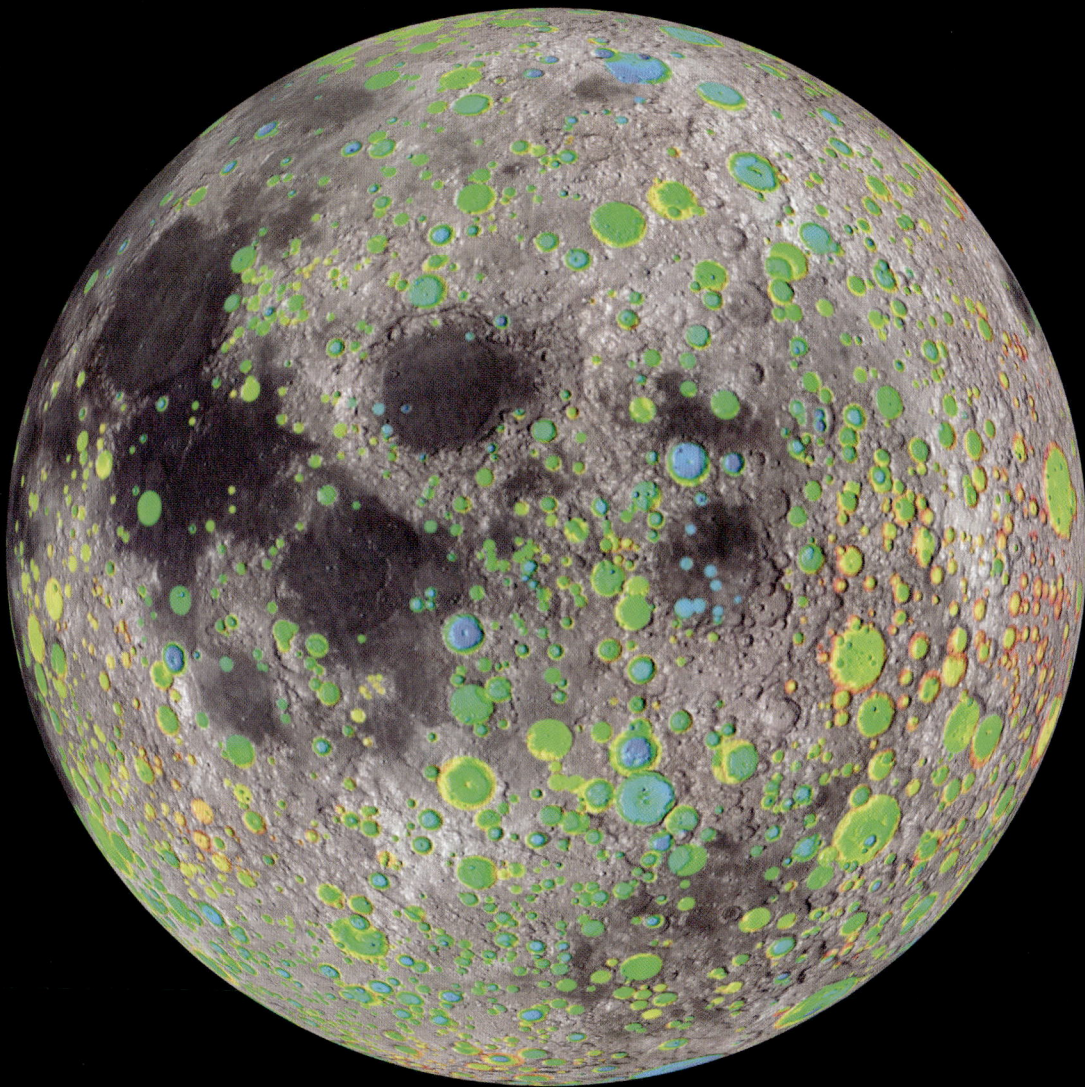

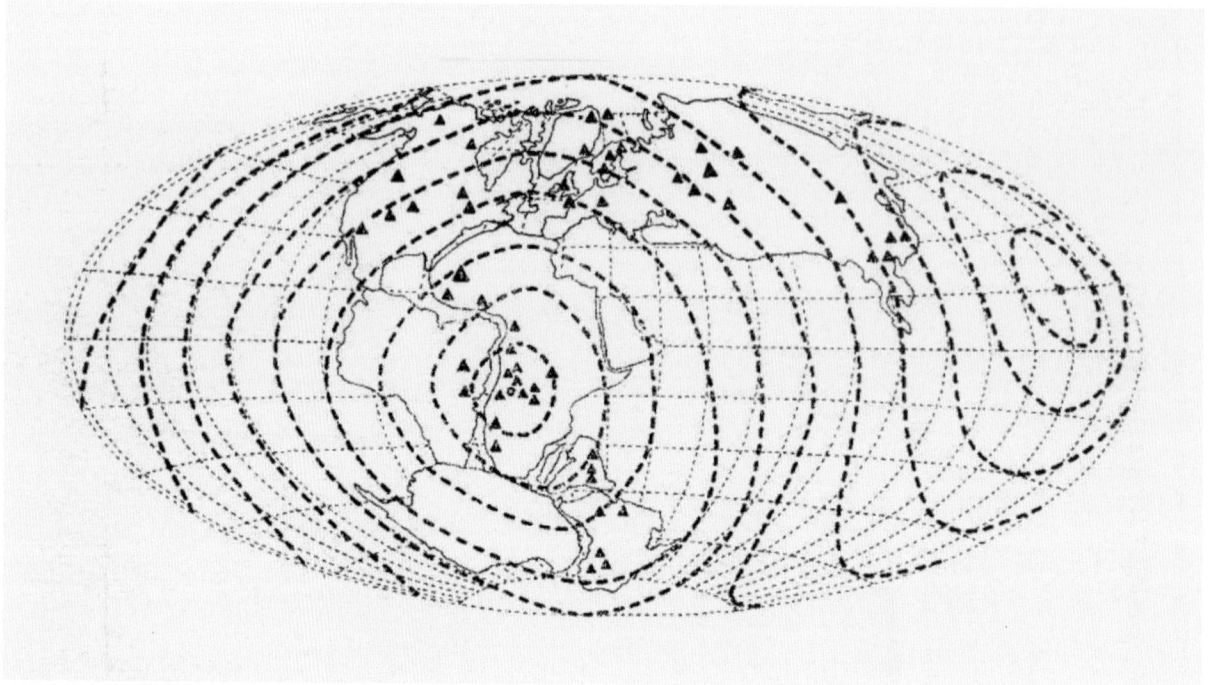

ABOVE W. B. Harland, illustration showing representative alleged Varangian tillites, from "Critical Evidence for a Great Infra-Cambrian Glaciation," *Geologische Rundschau*, vol. 54, pp. 45–61, 1964.

One of the most thought-provoking ideas regarding the Precambrian is "snowball Earth"—a theory first proposed in 1964. According to this idea, Earth went through two phases when it was entirely, or almost entirely, covered by glacial ice—720 to 660 million years ago and 650 to 530 million years ago. The evidence for this is that sediments deposited by glaciers ("tillites," marked by triangles in the image) dating from these ancient times have been found all around the world, including regions which were then located in the tropics. (The dashed lines indicate lines of latitude thought to prevail at that time.

This study was published when it was thought that the Earth's axis of rotation had wandered dramatically over time, so that when the tillites were deposited, one pole was off the coast of Brazil and the other was in the middle of the Pacific. We now know the situation to have been more complex.) Although many still support the idea that the Earth went through such a dramatic change in its climate, there are problems with the theory. First, it is difficult to be certain about the geographical location of the continents so far back in the past. Second, if the Earth did freeze over entirely, its dazzling surface would reflect so much solar radiation back into space that it seems unlikely a snowball Earth could ever thaw.

RIGHT Artist unknown, surface of the Moon in the region of Mare Crisium at New Moon, from "Astronomical Observations made at the Royal Observatory, Edinburgh," 1857.

An even more dramatic phase in Earth's history may have taken place much earlier in the Precambrian era. Patterns of cratering on the Moon's surface, along with ratios of uranium and lead isotopes measured in moon rocks returned by the Apollo missions, indicate that the inner solar system was exposed to a cataclysmic battery of asteroids and comets between 4,000 and 3,800 million years ago. Because this is some time after the formation of the solar system (4,600 million years ago) and the lunar crust (4,400 million), it is rather charmingly called the "Late Heavy Bombardment"—which shows how geologists' sense of time differs from the rest of us. Since the initial studies were published in 1974, further evidence has also been gathered from Mars, Mercury, and Callisto, a moon of Jupiter.

The Bombardment would surely also have affected Earth, and it is hard to imagine life starting until it was over. This image shows the Mare Crisium or "Sea of Crises," a large flat plain formed by volcanic basalts around 3,300 to 2,500 million years ago. Its relative lack of craters is evidence that the main phase of impacts was already over by the time of its formation.

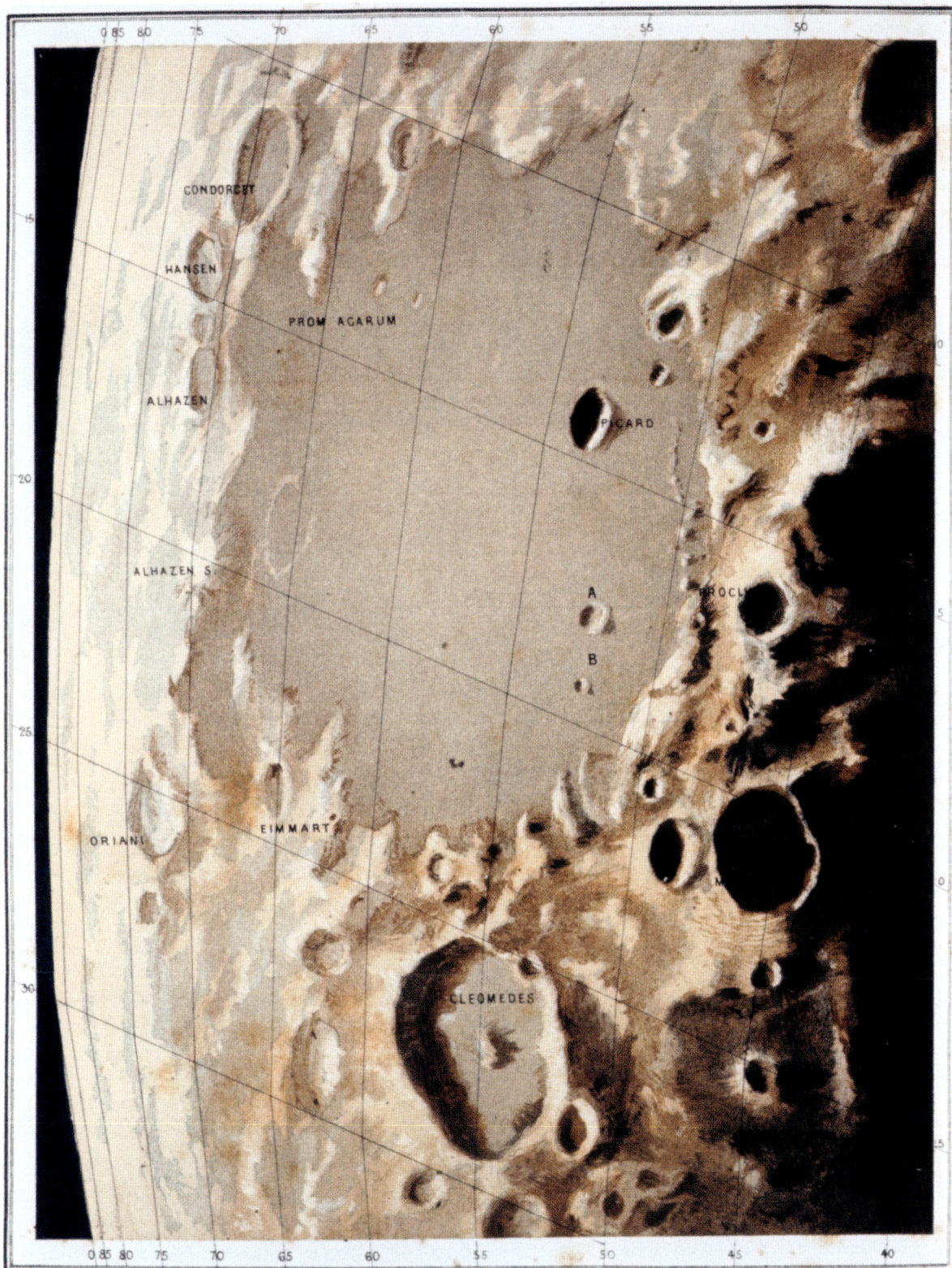

CONDORCET

HANSEN

PROM AGARUM

ALHAZEN

PICARD

ALHAZEN S.

A

B

ROCL

ORIANI

EIMMART

M

CLEOMEDES

Ancient of Ancients

FINDING EARTH'S OLDEST ROCKS

Now that radiometric dating has provided us with an accurate measurement of the age of the Earth—4,540 million years—this leaves a nagging question: what is the oldest rock that remains on the planet? And is that rock, as one might expect, the oldest object humans have encountered?

Isotope-based radiometric dating (page 45) suggests that the oldest known solid matter of Earth origin is zircon crystals discovered in the Jack Hills of far west Western Australia. Zircons are immensely durable crystals made of the elements zirconium, silicon, and oxygen in a 1:1:4 ratio (chemical formula $ZrSiO_4$) but occasionally zirconium atoms within the crystal structure are substituted by uranium atoms, and it is these that have allowed the crystals to be dated to 4,404 million years ago.

Many of the world's earliest mineral traces suggest that the planet was wetter and somewhat more like it is today than we might have thought. For example, other zircons indicate that the cores of today's continents were already forming, and a type of rock called gneiss up to 4,020 million years old is evidence of continental crust similar to what we see today. Rocks up to 4,310 million years old from the Nuvvuagittuq greenstone belt in northern Canada provide

direct evidence of underwater eruptions, so it seems oceans existed by that early time.

It is astounding that specimens as old as these have escaped destruction by the planet's continual geological activity. The Earth has a great ability to obliterate its own surface—consider, for example, what has happened to the thousands of craters it must have acquired during its history. Yet a few mineral relics have survived unchanged to the present day.

Recently it has been discovered that even older objects can be found on Earth, some even older than the Earth itself. On September 28, 1969, a large meteorite fell near Murchison in the Australian state of Victoria. Analysis of the meteorite showed that it contained irregular clumps of carborundum or silicon carbide (formula SiC) and using dating techniques involving helium and neon some of these were shown to be a staggering 7,000 million years old—thus pre-dating the solar system by more than 2,000 million years.

These presolar grains are indeed a vestige of a time before the Earth even existed. They were probably spewed out in the death throes of giant stars, and then drifted in space before adhering to some of the material forming our solar system. Had they landed on Earth, as presumably many did, these ancient particles would surely have been destroyed by its geological processes. However, those which became embedded in the Murchison object as it drifted through space were preserved right up until the moment it fell to Earth.

"The Earth has a great ability to obliterate its own surface."

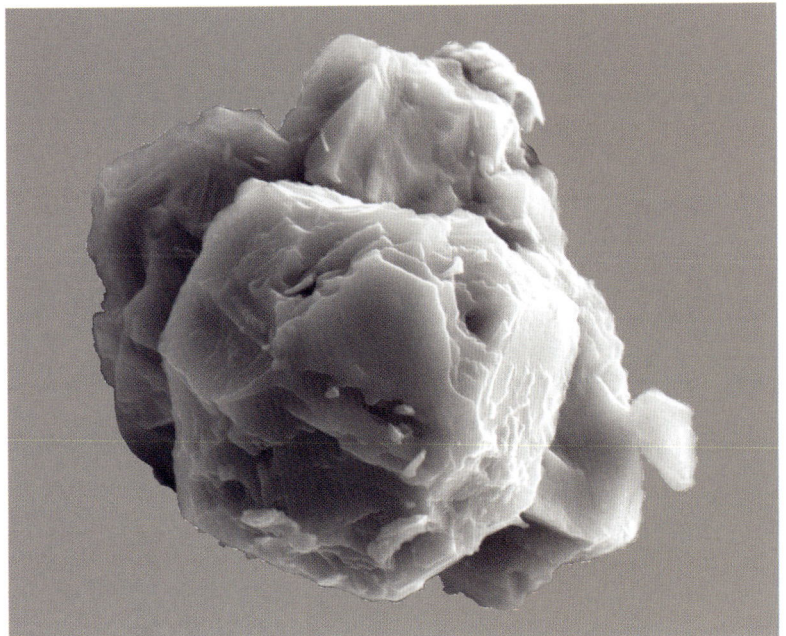

LEFT Acasta gneiss, northern Canada.

Some of the oldest known regions of the Earth's crust are in northern Canada, dated to 4,020 million years ago.

RIGHT Janaína N. Ávila and Phillip Heck, Presolar SiC grain, 2020.

This is an electron microscope image of a silicon carbide grain found on Earth, but whose age is estimated to be greater than that of our planet. Thus this tiny grain, measuring only eight thousandths of a millimeter, must be extraterrestrial and indeed pre-terrestrial in origin.

LEFT Dickinsonia fossil from the White Sea region of Russia, 2018.

Since the discovery that the Earth is older than the start of the Cambrian period 539 million years ago, geologists have tried to reach further and further back into what has, since perhaps 1888, informally been known as the Precambrian. This enormous expanse of time is now divided into the Proterozoic, Archaean, and the suitably ominous-sounding Hadean eons—and although it must have included the epochs when the Earth's surface became solid, the oceans formed, the continents appeared, life started, and the chemistry of the planet's atmosphere and rocks changed utterly, its sheer antiquity makes studying it extremely difficult.

This image is of *Dickinsonia*, a member of the Ediacaran biota, an assemblage of exotic multicellular organisms which flourished before more familiar forms appeared during the Cambrian. These fossils were discovered in the 1940s in South Australia by geologist Reg Sprigg (1919–1994). Initially his discovery was largely ignored, and only over the coming decades did the importance of these creatures become clear— indeed, biochemical evidence suggests that *Dickinsonia* was in fact an animal. The Ediacaran biota effectively rewrote the story of early multicellular life on Earth, and demonstrate how shifts in our understanding of deep time can still surprise us.

ABOVE OSIRIS-REx sample return capsule, 2023.

In the early morning of Sunday, September 24, 2023, the matt black sample return capsule of the OSIRIS-REx space probe touched down in the Utah desert. It contained material gathered from the asteroid 101955 Bennu, expected to revolutionize our understanding of the origins of organic molecules and life on Earth. Asteroids represent the debris left over from the formation of the solar system, and thus offer an opportunity to examine the very stuff from which our planet was made, preserved in arcing orbits since that ancient time.

2

ENERGY

Energy is what allows things to happen. In a physical system like the Earth it is energy that makes things move, rise, warm, bend, break, glow, and rumble. All these things have been happening on a large scale for a long time, so there must have been an enormous amount of energy inside the planet when it formed. And this energy is also the reason why it has settled into several layers, and why the continents continually crawl across its surface.

Kawah Ijen Volcano, Java, Indonesia, 2009.

One of the strangest natural phenomena in the world, the "blue lava" of the Javan volcano Kawah Ijen is not itself actually blue. Instead the blue color is caused by the burning of emitted gases as they mix with oxygen in the atmosphere.

> "Silence upon the whole subject; and let no one get before us in this design of discovering the center of the Earth."

Jules Verne, *Journey to the Center of the Earth*, 1864

The ancient Greeks realized there must be something *motive* inside the Earth to drive its sometimes-violent activity. Plato (427–348 BCE) thought it was permeated by rivers of vivifying fire, while others claimed its caverns echoed to the rushing of winds which powered volcanic eruptions. Later, the philosopher René Descartes (1596–1650) suggested the planet has a central fiery core, covered by several layers of dense matter, and its magnetic influence permeates it all. In his *Mundus Subterraneus* of 1664 the German monk Athanasius Kircher (1602–1680) proposed that the Earth is permeated by interlaced veins of water and fire, the former replenishing its rivers, the latter its volcanoes.

Even before the physics of heat were understood (we now know it is energy manifested as the movement of atoms within a substance) many thinkers realized that heat has seeped from the depths of the Earth since its existence. Lord Kelvin tried to calculate the age of the planet by estimating its rate of cooling (see page 40), but he was not the first to do so. The naturalist Georges-Louis Leclerc, Comte de Buffon (1707–1788), thought the planet formed as a hot, molten spinning ball and estimated that it must be at least seventy thousand years old by extrapolating from measurements of the cooling of red-hot iron spheres in a foundry. He believed life could only arise once the temperature had decreased to a temperate level, but also speculated that it would one day be extinguished when the world has chilled too much.

However, not all thinkers were convinced of the Earth's inner heat, nor that its center is molten. The German geologist Abraham Gottlob Werner (1749–1817) proposed that many of the rocks we see today crystallized out from a primordial global ocean—and he included in this rocks such as granite, which we now think of as igneous, formed

René Descartes (1596–1650), illustration of a magnetic field from *Principia Philosophiae*, 1644.

The philosopher René Descartes was one of the first to depict the Earth's magnetic field. He suggested it was created by moving particles which burrow through magnetic materials inside the planet, and are ejected from one pole in arcing loops which eventually return them to the other.

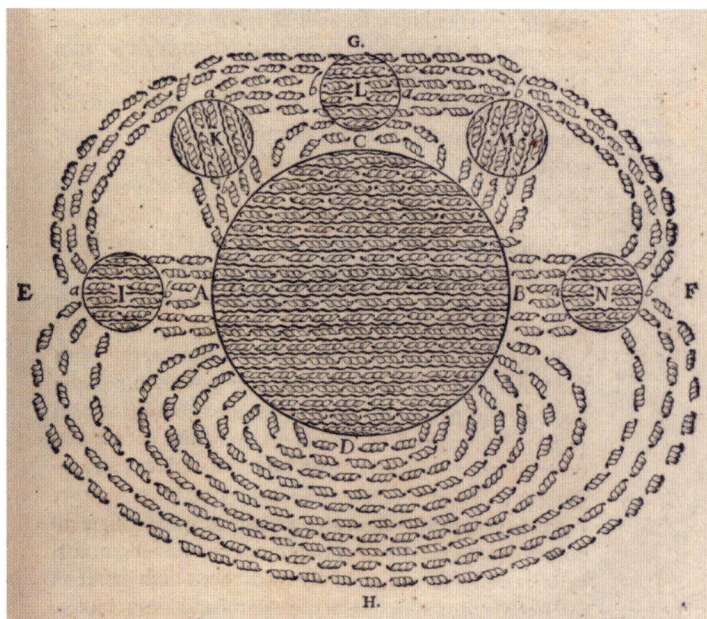

LEFT G. J. Symons, Plate 1 from *The Eruption of Krakatoa*, 1888.

Although not the largest nor most disruptive eruption of the nineteenth century, the 1883 self-demolition of Krakatoa in the Dutch East Indes was to become certainly the most famous.

occurred—the eruption of the Indonesian volcanoes Tambora in 1815 and Krakatoa in 1883, and Martinique's Pelée in 1902, as well as the earthquakes in Arica, Chile, in 1868, and San Francisco in 1906. Eventually Hutton's Plutonism was to prevail, and other scientific discoveries of the era slotted into place. Through taking measurements in mines at various depths, professor of geology at the Muséum national d'Histoire naturelle in Paris, Louis Cordier, discovered that, in general, the ground beneath our feet becomes hotter by 1°C (1.8°F) for every 45 m (150 ft) one descends. He extrapolated this further to predict that approximately 100 km (60 mi.) down the Earth must become molten. In 1903, the question of where this heat comes from was answered by Ernest Rutherford (1871–1937) and H. T. Barnes (1873–1950) who showed that radioactive decay releases heat energy. One fundamental problem still remained—it seemed impossible to peer directly into the Earth's internal workings.

This all changed in 1906 with the discovery by the British geologist Richard Dixon Oldham (1858–1936) that earthquakes generate different

RIGHT The Earth's geothermal gradient.

One notable feature of the geothermal gradient is how rapidly temperature rises within the thickness of the crust. This is thought to be because half of all heat energy generated by radioactivity within the Earth is produced in its crust—despite its relative thinness. The temperature rises slowly across the mantle and outer core because they are viscous liquids which dissipate heat "upwards" by convection. As a result, the very center of the Earth may be at "only" 4,700–6,000°C (8,500–10,800°F).

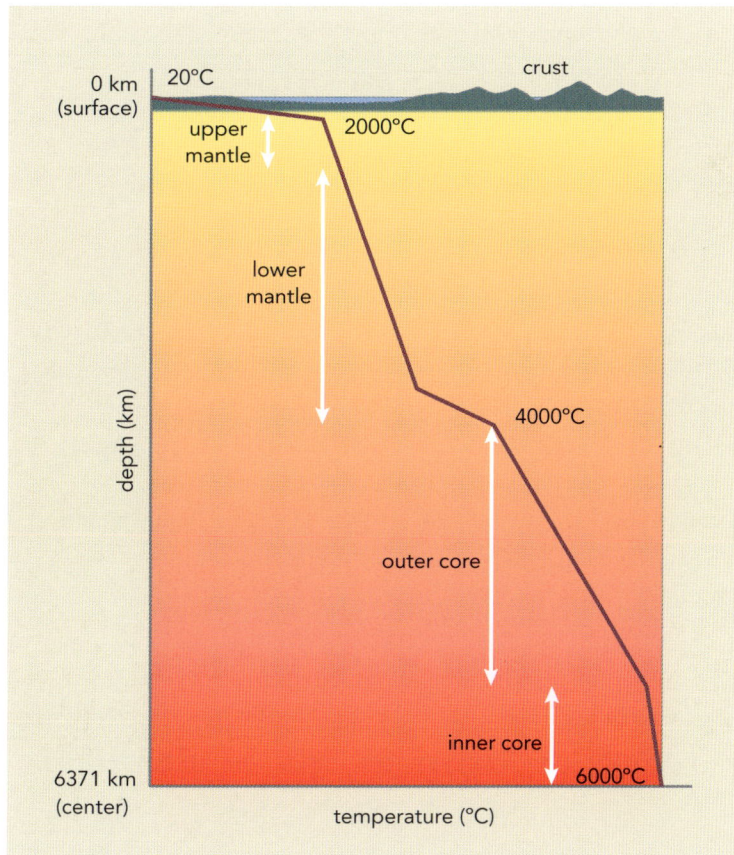

types of seismic wave, which propagate through the planet in ways that suggest it is divided into two internal layers, the mantle and the core. The radius of the core is approximately 3,500 km (2,200 mi.), while the mantle extends to within tens of kilometers of the solid surface (which lies at a radius of 6,370 km [3,960 mi.]). We have no samples of the core but we know it is much denser than the mantle. We do, however, have mantle specimens and they are chemically similar to the solid crust. The discovery of these layers of different density showed that the constituents of the planet have settled over time under the influence of gravity, so its center must have been liquid at some point. Indeed, we now believe that the Earth's mass is still mostly comprised of liquids, even if some of those liquids are extremely viscous.

German postage stamp commemorating the theory of continental drift popularized by the German meteorologist Alfred Wegener (1880–1930), 1980.

Also, in those early years of the twentieth century, scientists' attention was drawn to a different aspect of the globe's internal fluidity—continental drift. The idea that the Earth's land masses slowly migrate dates back at least as far as Robert Hooke in the seventeenth century. Since then, scientists including Alexander von Humboldt (1769–1859) and Antonio Snider-Pellegrini (1802–1885) had noted that some continents seem to have complementary shapes, as if they once fitted snugly together— western Africa and eastern South America are a famous example.

Continental drift reached its most clear expression in the work of the German meteorologist Alfred Wegener (1880–1930). In his 1915 *Die Entstehung der Kontinente und Ozeane* (*The Origin of Continents and Oceans*), he argued that today's continents were once fused into a single *Urkontinent* or *Pangäa* (Pangaea as it is called today), which fragmented to allow the continental land masses to drift apart and attain their present positions. Unfortunately, Wegener wrongly suggested that continental drift was caused by changes in the Earth's rotation, and the phenomenon was to remain controversial, lacking a convincing underlying mechanism for the next five decades.

"In 1964 it was shown that the fit between the continents is even more convincing if one tries to match up their offshore continetnal shelves, where the continental crust suddenly drops away to the deep ocean floor."

Over those decades, evidence for that mechanism gradually accumulated. It had been known for some time that seismic and volcanic activity is focused along sinuous lines that arc across the surface of the globe, the Pacific "Ring of Fire" being the best example. In the 1950s a different planet-circling network was clearly charted for the first time—an interlinked web of long mid-ocean ridges which sweep through the Atlantic, Pacific, Indian, and Southern Oceans. Soon after, it was realized that magnetic stripes in seafloor igneous basalts can be used to track the origin and movements of oceanic crust over time. Finally in 1964 it was shown that the fit between the continents is even more convincing if one tries to match up their offshore continental shelves, where the continental crust suddenly drops away to the deep ocean floor.

These ideas coalesced at the 1965 Royal Society of London Symposium on Continental Drift where geologists from around the world established

Michael Wutky (1739–1823), *The Phlegraean Fields*, c. 1780.

The Phlegraean Fields is an active volcanic caldera on the northern edge of the bay of Naples, and part of the same volcanic complex as Vesuvius. The terrain here continually rises and falls over the centuries, which was important in the discovery that the planet changes over different timescales—some imaginably long, some within the scope of human history.

plate tectonics as the mechanism which underlies continental drift (tectonic means building). According to this theory, Earth's crust is made up of multiple raft-like plates which drift independently, floating on the mantle below. The mid-ocean ridges are places where new seabed is continually created, forcing the existing ocean crust apart on either side of the ridge, and thus pushing continents away from each other (as occurs in the Atlantic). If some oceans are getting wider then others must be getting narrower, and in much of the Pacific the plates of oceanic crust are being forced under other plates to be lost into the underlying mantle—at so-called "subduction" zones.

Plate tectonics explains so much about Earth that it has become the central theory of geology. Ocean crust is made and destroyed all the time, so most is younger than 100 million years, whereas most continental crust is over 3,500 million years old. The energy released at subduction zones is the cause of most of the world's earthquakes and vulcanism, and most mountains are formed where plates collide and crumple. Today's continents

Artist unknown, "A Prospect of Mount Aetna, with its Irruption in 1669," date unknown.

The tallest active volcano in Europe, Mount Etna dominates the northeastern corner of the island of Sicily. Frequently disgorging smoke and recasting its summit with eruptions, it has in the past threatened nearby towns.

Thomas Moran (1837–1926), *The Great Blue Spring of the Lower Geyser Basin, Yellowstone National Park,* 1873.

Yellowstone National Park straddles one of the world's largest supervolcanoes—where a plume of hot material from deep within the planet reaches the surface to power geysers, hot springs, and the occasional cataclysmic eruption.

were once joined together, explaining the presence of related animal and plant species in regions now separated by thousands of miles of ocean. The list is a long one.

All this tectonic activity requires vast amounts of energy, and the 1965 tectonics pioneers suggested this comes from the rolling motion of the fluids in the viscous liquid mantle—convection like that seen in a pan of boiling water. However, it remains unclear exactly how the roiling of the mantle actually causes the movement of the tectonic plates above. Indeed, the energy systems of the Earth still hold many mysteries for today's geologists. How does energy flow between the different layers of the Earth? How are the regions of the mantle arranged? Why is there more igneous granite in the world than we would expect? How did the cores of continents first appear? What started the Earth's tectonic activity, and when? Finally, why do we not see evidence of tectonic activity elsewhere in the solar system?

Energy is central, essential in fact, to all geological processes, but there is much more geological discovery still to come.

ABOVE **Model of Zhang Heng's seismometer of 132 CE.**

The first known geological instrument was created by the Eastern Han dynasty polymath Zhang Heng (78–139 CE). It was designed as a seismometer to measure the direction in which the ground shakes during earthquakes (technically, *seismometers* measure seismic activity, but can be coupled to *seismographs*, which create some sort of recording). The vase is fixed to a firm base on the ground, whereas a heavy internal pendulum inside it can move more freely and thus lag behind due to its inertia when a quake pushes the vase sideways. The pendulum hits a rod that triggers one of the ornate dragons on the vase to spit a brass ball into the mouth of a waiting frog. If several detectors in different locations are triggered, the direction whence the seismic waves reached each detector can be traced back to determine the location of the earthquake's epicenter—the point on the Earth's surface directly above where the earthquake originated.

LEFT Artist unknown, *Battling the Namazu Catfish*, c. 1855.

Due to the country's underlying geology, much lore relating to Earth's more energetic processes are derived from Japan. One which unfortunately has not survived to the present day is the myth of the Namazu, the giant subterranean catfish whose wriggling is the cause of earthquakes. It is said that a fisherman saw an unusually active catfish the day before the 1855 Edo earthquake, and that this is the source of the story—although it is likely that the tale of the seismic catfish is considerably older.

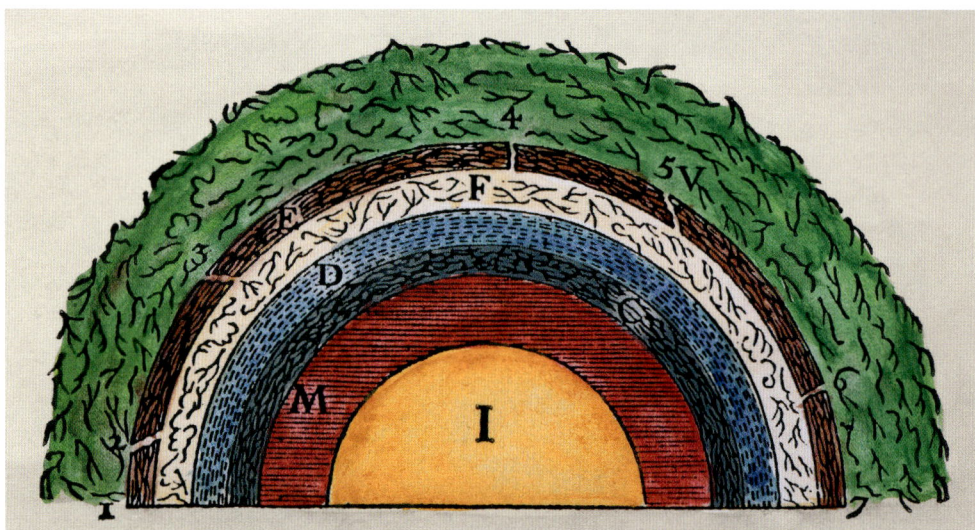

René Descartes (1596–1650), diagram from *Principia Philosophiae*, 1644.

The renowned philosopher René Descartes had a flawed view of physics, one which did not survive long after the revolution led by Isaac Newton (1642–1726). According to his cosmology, the universe contains three elements: fire, water, and earth. There is no vacuum around Earth so its constituents are hemmed in by the celestial matter by which it is surrounded—and this is what explains the tendency for heavy things to fall toward the center of the planet under the influence of what we would now call gravity.

However, Descartes did propose a system by which the constituents of the globe have settled into a series of layers,

partly according to their weight—layers of air, water, crust, and metals, with an inner fire. This is a model not dissimilar to what exists today, in which Earth is thought to have gone through a process of "planetary differentiation." Energy is key to this: planets must be at least partially molten for materials to swap places within it. Also, differentiation is still ongoing today, and the energy it releases explains phenomena such as the Earth's magnetic field.

In this diagram Descartes expounds his ideas about the early formation of the Earth. In the upper image the planet has settled into a series of layers, but as it cools and shrinks, the crust cracks, forming in the lower part of the illustration a crumpled surface of mountains ("F") and depressions in which seas accumulate ("D").

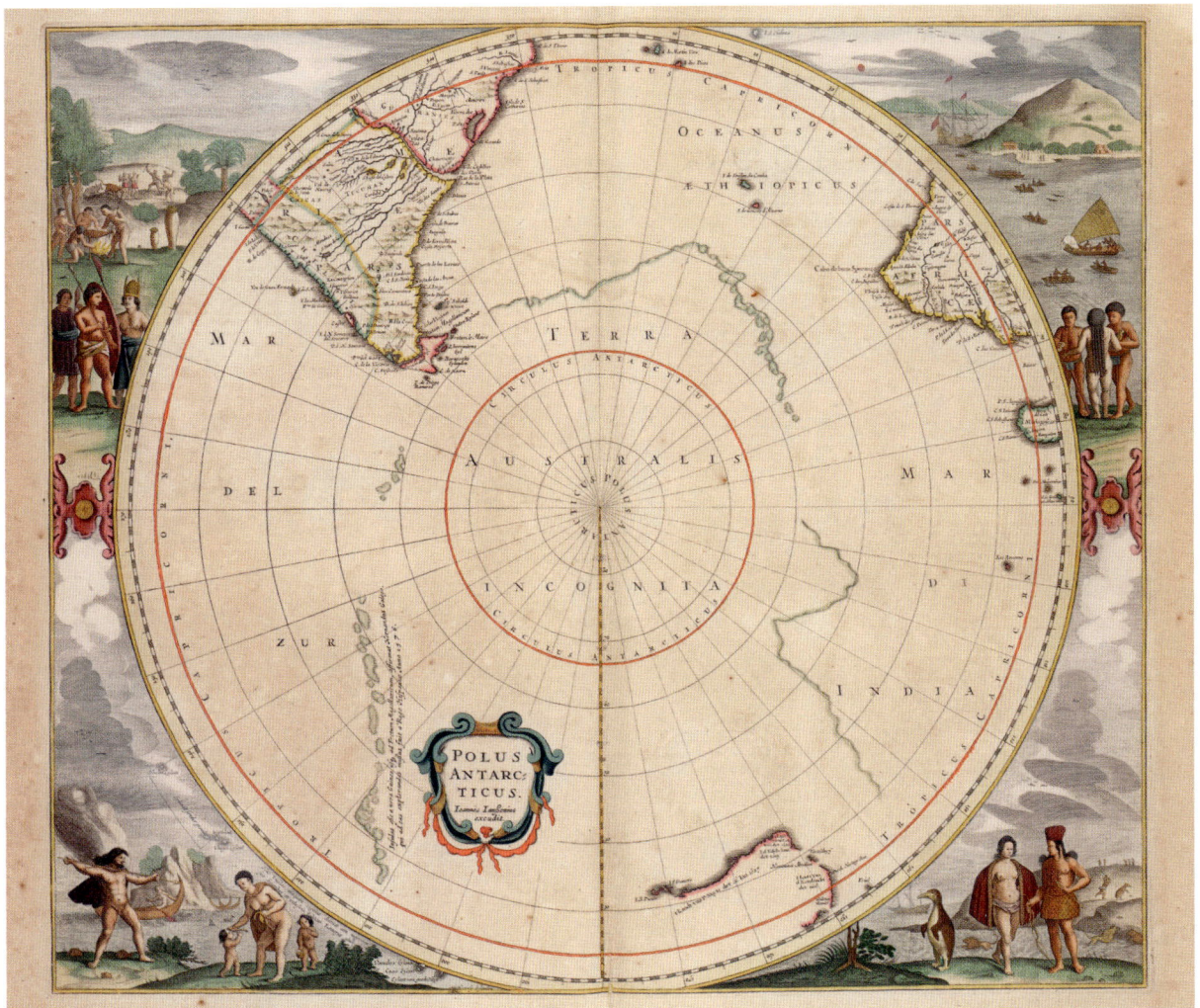

Jan Janssonius (1588–1664), "Polus Antarcticus cum regionibus subjacentibus et maribus illum alluentibus" (The Antarctic Pole with the underlying regions and the seas which flow into it), from the *Atlas Maior*, 1657.

Early conceptions of the energy balance of the Earth led to some surprising conclusions. Since the ancient Greeks, it had long been wondered why most of the world's known landmasses are in the northern hemisphere. It was concluded that a large unknown southern continent must exist to counterbalance the weight of the northern lands—the hypothesized Terra Australis Incognita. This landmass was duly inserted into many maps, and when Europeans first viewed the shores of Australia and Antarctica the theory must have seemed vindicated.

However, we now know there is no particular reason why the land should be distributed evenly—roughly two thirds of land is in the northern hemisphere, and around 220 million years ago almost all of it was aggregated into one continent, Pangaea (page 81). A similarly extreme example of global asymmetry is modern-day Mars, whose northern half is lower and smoother, while the southern is higher and pock-marked by impacts and volcanic activity.

Artist unknown, *Destruction de Lisbon*, 1755.

Just before 10 a.m. on November 1, 1755, three surges of seismic shaking hit the Portuguese capital of Lisbon over a period of approximately eight minutes, causing the most destructive European earthquake of recent centuries. The quake was followed an hour later by tsunamis which swept in from the River Tagus—the city is on the river's north bank, near its opening to the sea. The quake, the inundation, and the subsequent fire left many areas of the city in ruins, and evidence of its impact is still visible in this most beautiful of capital cities. The staggering power of the earthquake—damage occurred as far away as Seville in Spain and even in Morocco—spurred a new age of scientific enquiry into the inner workings of the Earth.

J.M.W. Turner (1775–1851), *Sunset*, c. 1830.

Peaking in April 1815, the eruption of Mount Tambora on the island of Sumbawa in what is now Indonesia, was in some ways the volcanic counterpart of Lisbon. It is the largest historically recorded volcanic explosion, blasting 50 cubic km (12 cubic miles) of rock from the ground, forming plumes up to 40 km (25 mi.) high, and blasting hot gases, debris, and water across the area. It injected a mist of sulfur dioxide (SO_2) into the stratosphere which cooled the planet for years, leading to crop failures, disease epidemics, and waves of human migration. In fact, artificial stratospheric aerosols have now been suggested as a solution to global warming.

A more innocuous effect of Tambora was to create stunning sunsets around the world over the following years. One of Britain's most renowned artists, J.M.W. Turner, was wont to paint sunsets and sunrises even in more atmospherically quiescent times, but in the two years after Tambora he produced a series of striking works depicting the reds, oranges, and purples the volcano's debris created in British skies.

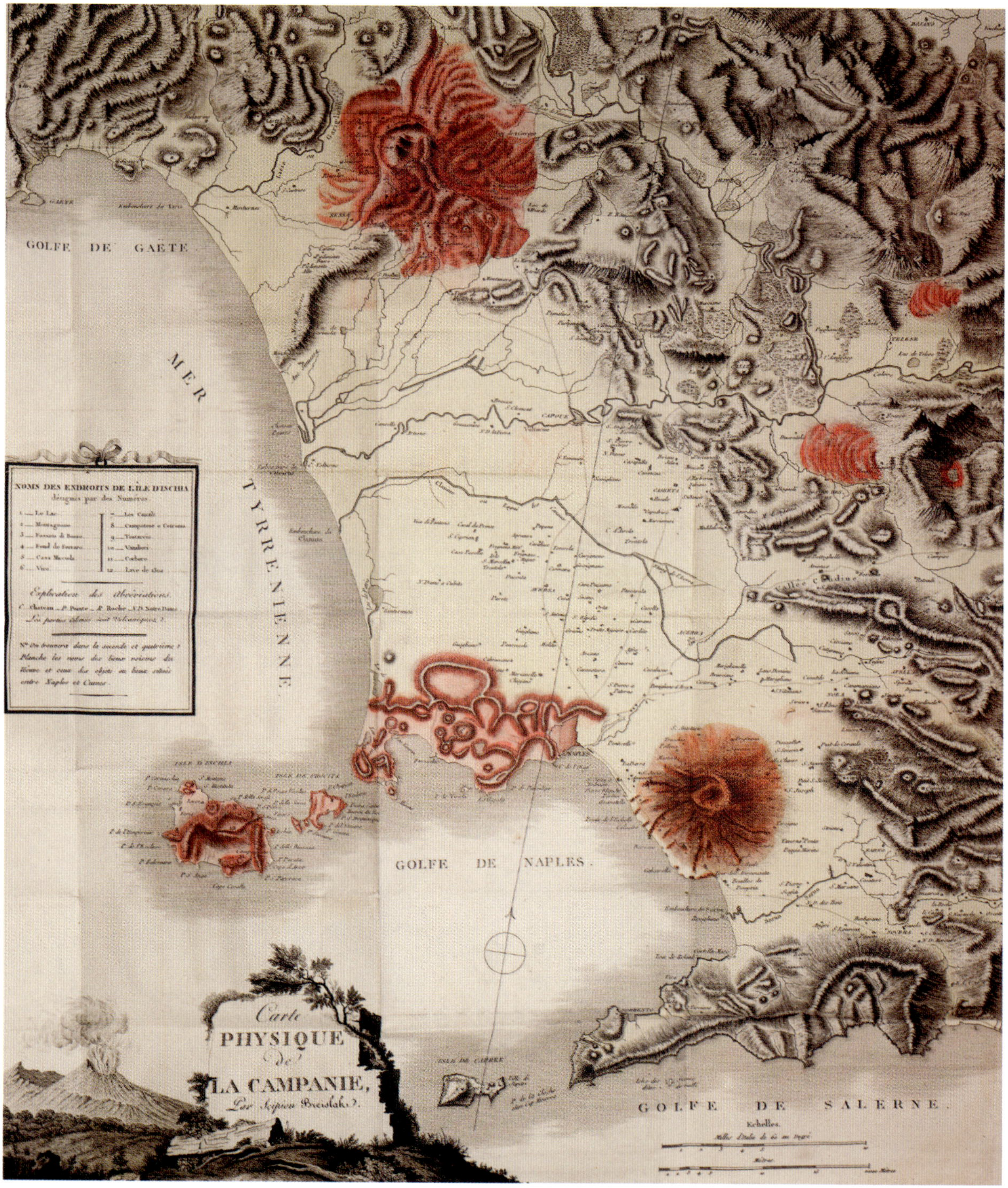
GOLFE DE GAËTE
MER TYRRENIENNE
NOMS DES ENDROITS DE L'ILE D'ISCHIA
désignés par des Numéros
Explication des abréviations.
C. Chateau _ P. Pointe _ R. Roche _ N.D. Notre Dame
ISLE D'ISCHIA
ISLE DE PROCITA
NAPLES
CAPOUE
GOLFE DE NAPLES
ISLE DE CAPREE
GOLFE DE SALERNE
Echelles
Carte
PHYSIQUE
de
LA CAMPANIE,
Par Scipion Breislak.

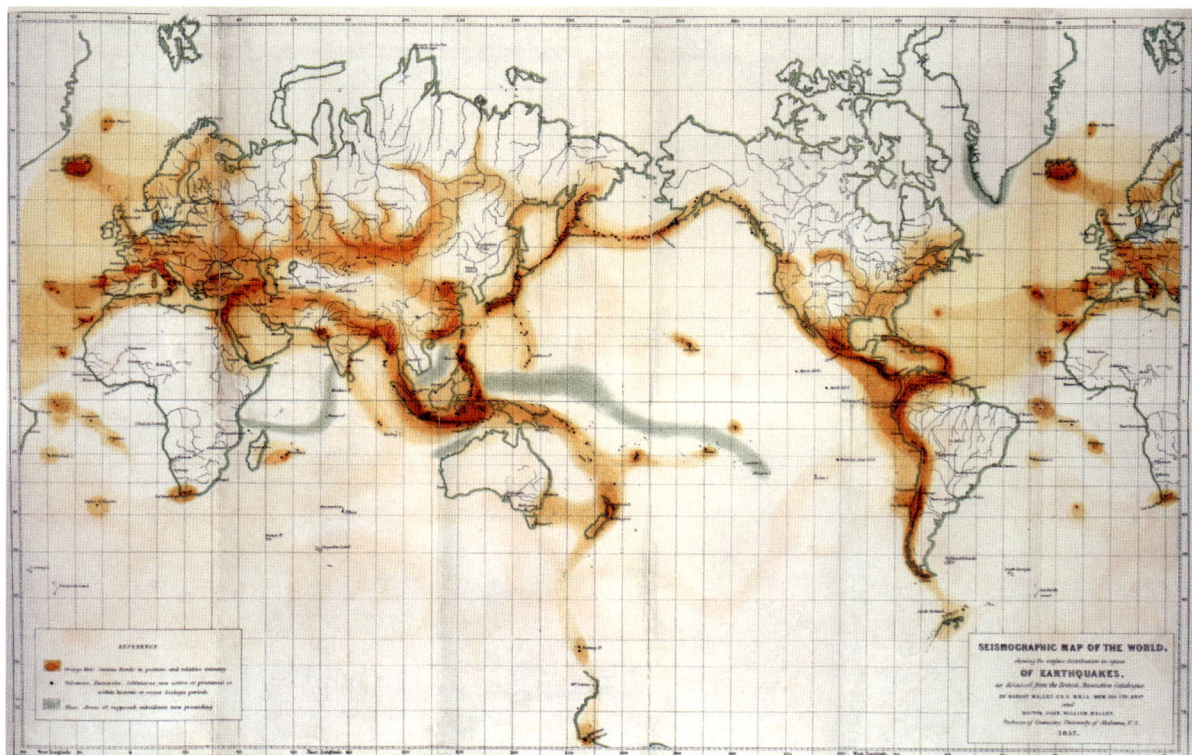

LEFT **Scipione Breislak (1748–1826), "Carte Physique de la Campanie," from *Voyages Physiques et Lythologiques dans la Campanie,* 1801.**

Of Swedish extraction, the geologist Scipione Breislak (1748–1826) worked mainly in Italy during a time when Western Europe was in the throes of great political upheaval. He was professor of physics at Ragusa in Sicily before a stint teaching in Paris followed by a return to northern Italy. This map dates from his stay at Naples and depicts the contorted topography of the northwestern part of the region of Campania. It shows unusual tangles around the Phlegraean Fields volcano (pages 32–33, 63) just below the center of the map and north of the Bay of Naples, as well as the serrated cone of Vesuvius immediately to the east.

ABOVE **Robert Mallet (1810–1881), *Seismographic Map of the World,* 1857.**

Some images look far too modern for the time they were created, and this seismographic map of the world by the Irish engineer and geologist Robert Mallet is just such an image. Centered, appropriately, on the Pacific, it makes clear that there is a pattern to the Earth's seismic activity—it is concentrated along sinuous and arcing lines, many of which follow obvious geographical features such as coastlines and mountain arcs.

Mallet coined the terms "seismology," "epicenter," and "seismic focus" (the point below the epicenter where an earthquake actually originates) and he used multiple lines of evidence to ascertain where individual quakes had occurred. He even proposed that volcanism is driven by energy derived from deformation within the Earth's crust. His publications have evocative titles, including *Report of the Facts of Earthquake Phaenomena* and *Volcanic Energy: An Attempt to Develop its True Origin and Cosmical Relations.*

Beneath the beauty of the Yellowstone caldera

HOW A SUPERVOLCANO BECAME THE FIRST NATIONAL PARK

Traveling from the Beartooth Highway into the northeast of Yellowstone National Park is to step back in time to a North America before the coming of Europeans a few short centuries ago. Rivers run untrammeled through tree-lined meadows, wolves and elk pursue their ancient war of attrition, and the seasons turn with brutal severity. And all this even before one passes the Grand Canyon of the Yellowstone River to reach the park's volcanic center.

Yellowstone has been inhabited for at least ten thousand years, and echoes to myths and legends of the steamy wonders which lie at its heart. Visits by European settlers were limited to the wanderings of occasional fur trappers, bringing back rumors of lakes hot enough to cook their own fish. The region was one of the last parts of the United States to be mapped. Although small expeditions passed through in the 1860s, many of their descriptions were simply not believed—and it is possible that the area's geothermal activity was even more spectacular then than now.

It took the expedition of geologist Ferdinand Hayden (1829–1887) to bring the wonders of Yellowstone to the outside world. Hayden's group mapped the area's geology and geography and, importantly, he took with him the photographer William Henry Jackson (1843–1942) and the artist Thomas Moran (1837–1926). Within a year there were calls for Yellowstone to be designated the first national park in the nation and, arguably, the world.

Yellowstone is centered on a vast caldera, a shallow 50-by-70-km bowl (30 by 45 mi.) fixed over a supervolcano—a depression which has slumped down over a vast dome of partially melted magma. Areas of the caldera rise or fall several centimeters a year, and the subterranean dome provides energy for the region's geothermal wonders. There are mudpots and bubbling lakes, hot springs colored by extremophile organisms, and more than half of the world's geysers.

It is now thought Yellowstone lies over a vast rising plume in the Earth's mantle, which may derive from as deep as its boundary with the core—conveying energy toward the surface and focusing it on one small region. In the last few million years this has resulted in enormous eruptions, probably three of them, blasting out between two and twenty times as much material as Tambora did in 1815 (page 71) and covering the continent west of the Mississippi with ash. On a longer timescale, it seems that North America has crept westward over the plume, and its trail of volcanic destruction has left behind the flat lands of the Snake River Plain.

Similar mantle plumes exist elsewhere on Earth—Hawaii's line of islands results from the Pacific seabed sliding diagonally over a plume, and Iceland is a unique example of a plume that happens to lie under a midocean ridge (page 89).

LEFT William Henry Jackson (1843–1942), *Mammoth Hot Springs, Yellowstone*, 1871.

RIGHT Thomas Moran (1837–1926), *The Castle Geyser, Firehole River, Yellowstone*, 1872.

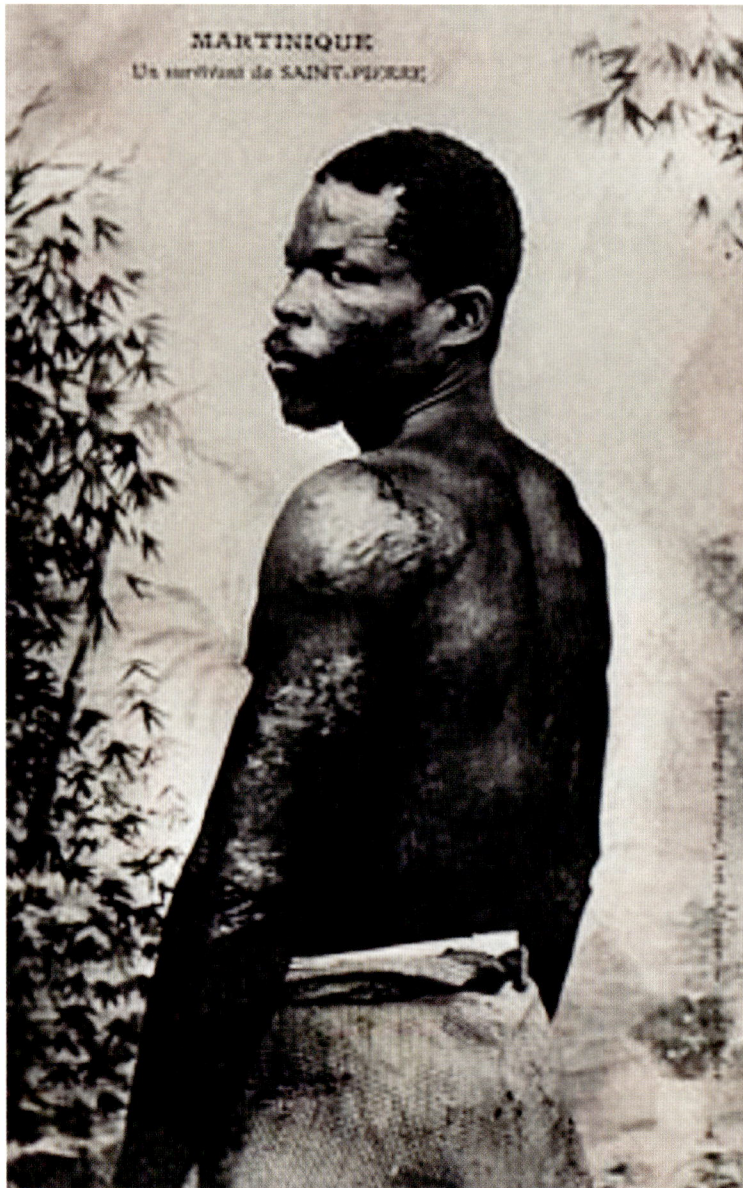

Unknown photographer, photograph of Ludger Sylbaris (1874–1929), 1902.

On the island of Martinique stands a stone prison cell with an arched roof, famed as the temporary residence of Ludger Sylbaris, where he survived the eruption of nearby Mount Pelée in 1902. Sylbaris had a checkered past of drinking and crime, and had been thrown into the tiny, cramped cell by the local police, a punishment which saved his life. Within seconds of the main eruption of the volcano, a burning cloud of gas and dust immolated or suffocated perhaps thirty thousand people within 12 km (7 mi.) of the peak. The prisoner in the stone cell survived, albeit with severe burns—"the man who lived through Doomsday."

Unknown photographer, Union Streetcar Line after the 1906 San Francisco Earthquake, looking west.

Over a period of approximately one minute just after 5 a.m. on April 18, 1906, the two sides of the San Andreas fault suddenly slipped past each other over a length of almost 500 km (300 mi.), causing one of the most devastating earthquakes of the twentieth century. The epicenter was located near San Francisco and destruction was worst in areas built over layers of sediment, or land reclaimed from the Bay. Untold hundreds died, and contact was lost with the outside world—indeed, many feared that humanity had been struck by a global catastrophe. Later, the configuration of the quake led geologists to propose that earthquakes occur when energy builds up in rocks over many years, as they are elastically deformed by movements of adjacent regions of crust, and then is suddenly released.

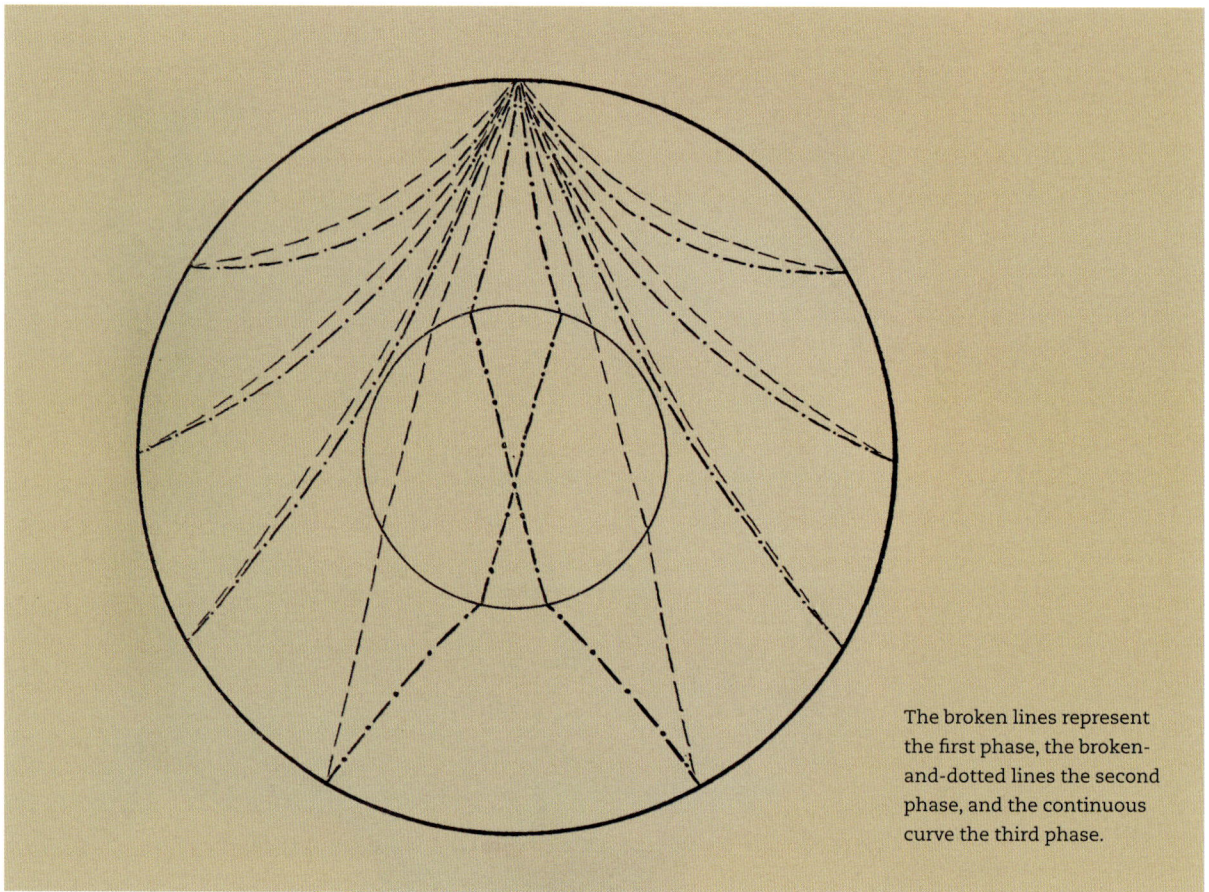

The broken lines represent the first phase, the broken-and-dotted lines the second phase, and the continuous curve the third phase.

Richard Dixon Oldham (1858–1936), illustration from "The Constitution of the Interior of the Earth, as Revealed by Earthquakes," *Quarterly Journal of the Geological Society,* vol. 62, pp. 456–75, 1905.

Richard Dixon Oldham was the first person who allowed us to peer deep inside the structure of the Earth, and finally detect an arrangement somewhat similar to Descartes' layers (page 68). Working for the Geological Survey of India, he identified the different kinds of seismic wave produced by earthquakes, and used them to discover the planet's layered structure.

There are three types of seismic wave with different properties. P-waves are pressure waves, in which matter within the Earth throbs in a direction parallel to the direction the wave is traveling—they move the fastest and can go through solids and liquids. S-waves are transverse waves that make rock vibrate in a direction perpendicular to the direction the wave is traveling—they can go through solids, but only the most viscous liquids. Finally surface waves occur when the ground itself sways up and down—and these are the slowest waves of all.

As seismic waves pass deep into the planet, the increasing density tends to make them slow down, although this effect is outweighed by the fact that increasing rigidity tends to speed them up. The net effect is that seismic waves within the Earth often follow curved trajectories, bending gradually toward the surface. In Oldham's diagram above, waves radiating from a single point on the Earth at the top of the diagram are seen curving in this way. In addition, seismic waves suddenly divert when they cross from one layer to another – and in the diagram this can be seen at the boundary between Oldham's proposed core and surrounding mantle layers.

Careful measurement of the arrival of seismic waves around the world has now allowed us to discover the different layers below our feet. The solid crust transitions into the mantle in a complex manner at a depth of 10–75 km (6–47 mi.) where seismic waves suddenly speed up—the Mohorovičić discontinuity discovered in 1909. Much deeper, at 2,900 km (1,800 mi.), lies the boundary between the mantle and the core—a sphere of much denser material, mainly metallic iron and nickel.

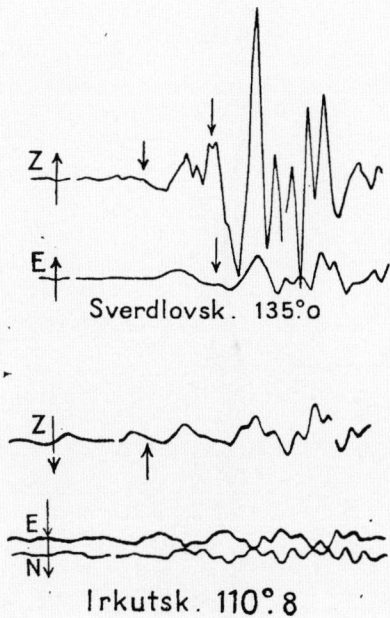

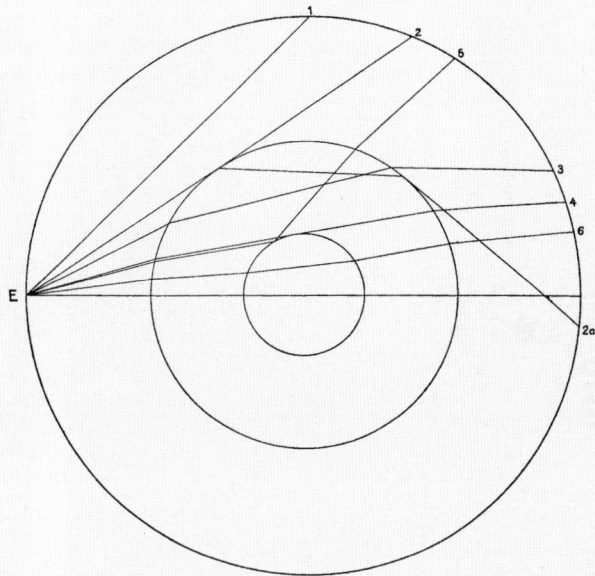

Inge Lehmann (1888–1993), "Paths Through the Earth," illustration from the essay "P'," *Publications du Bureau Central Scientifiques*, vol. 14, pp. 87–115, 1935.

Even before Oldham detected the boundary between the mantle and the core (see opposite), it had been assumed that the Earth must have a dense center. On average the planet was calculated to have a mass of approximately 5,500 kg for every cubic meter (310 lb per cubic ft) of its volume, whereas the average density of most rocks, including those in the vast mantle, seems unlikely to exceed 3,500 kg per cubic meter (220 lb per cubic ft). However, a vast, spherical, dense metal core at the center of the Earth neatly explains this discrepancy—the density of iron at the earth's surface, for example, is 7,874 kg per cubic meter (491 lb per cubic ft).

However, the seismologist Inge Lehmann, working at the Danish Geodetic Institute, worried that existing models of the Earth's layers did not quite fit the seismological data—

in short, she realized that, after earthquakes, P-waves arrive in places where they should not arrive if the core were entirely liquid. A gifted mathematician, Lehmann calculated that this inconsistency could be explained if the core comprises two parts—an outer liquid core and an inner solid core, separated by a boundary now called the Lehmann discontinuity in her honor. These two core layers are the two internal circles in the image above.

We now believe the outer liquid core has a radius of 3,500 km (2,200 mi.), and the inner solid one 1,300 km (800 mi.). At the mantle/core interface seismic waves slow down and are deviated toward the Earth's center, whereas at the outer core/inner core boundary they speed up. It is now also clear that heat loss from the inner core provides energy which induces enormous swirling currents within the liquid outer core—and this churning iron alloy generates the Earth's vast magnetic field, extending far out into space.

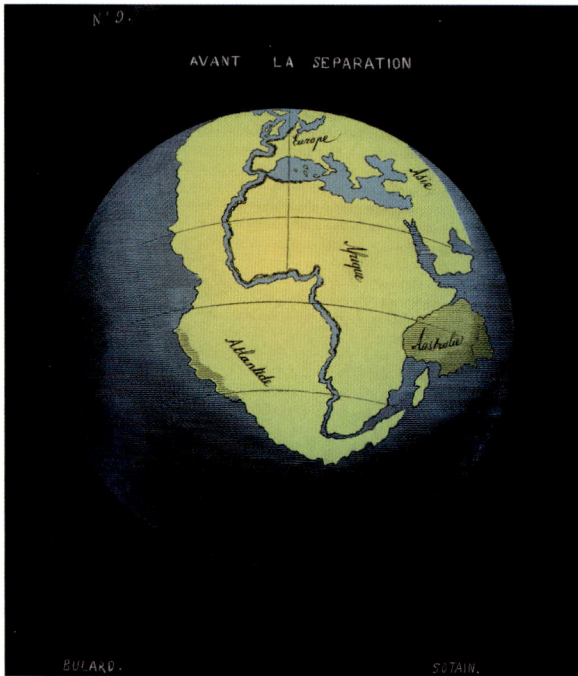

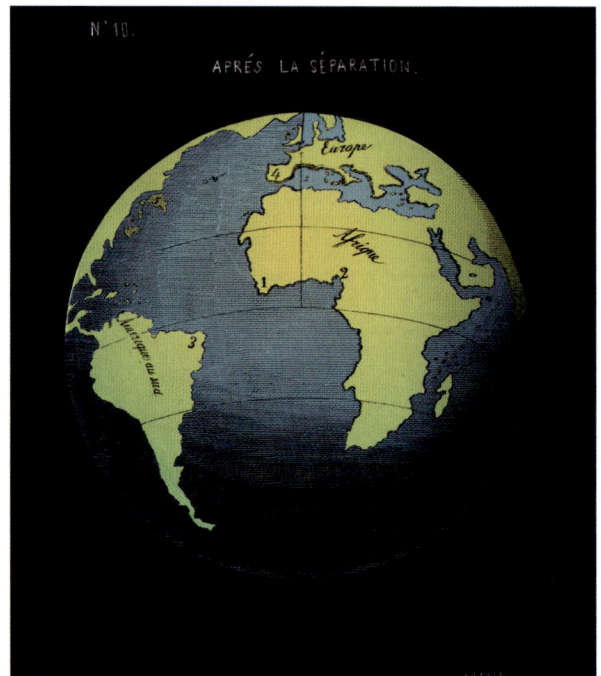

Antonio Snider-Pellegrini (1802–1885), "Avant la séparation" and "Après la séparation," from *La Création et ses mystères dévoilés* (The Creation and Its Mysteries Unveiled), 1859.

The idea that the continents might creep across the surface of the globe, and even have split apart from each other in the distant past, is an old one. The first clear sketch of such a split was published in 1858 by the French geographer Antonio Snider-Pellegrini. He drew particular attention to the striking correspondence between the coastlines of eastern South America and western Africa and, presciently, even suggested that all the continents had once been fused into a single giant landmass.

These ideas, now called "continental drift," were resurrected in more complete form half a century later by the German meteorologist-astronomer-geophysicist Alfred Wegener, although he preferred to call the phenomenon "displacement theory." In a public talk on January 6, 1912, "The Formation of Large Features of Earth's Crust (Continents and Oceans) Explained on a Geophysical Basis," he laid out his evidence for continental drift, its implications, and its causes.

Inspired by the "jigsaw" nature of once-interlocking continents, the separation of geological features and related creatures between now-distant continents, fossilized tropical plants at the poles, and evidence of glaciation in Africa, he argued that the distribution of the continents had changed over time. He also proposed, correctly, that today's continents are the fractured remnants of an ancient uber-continent, Pangäa (Pangaea).

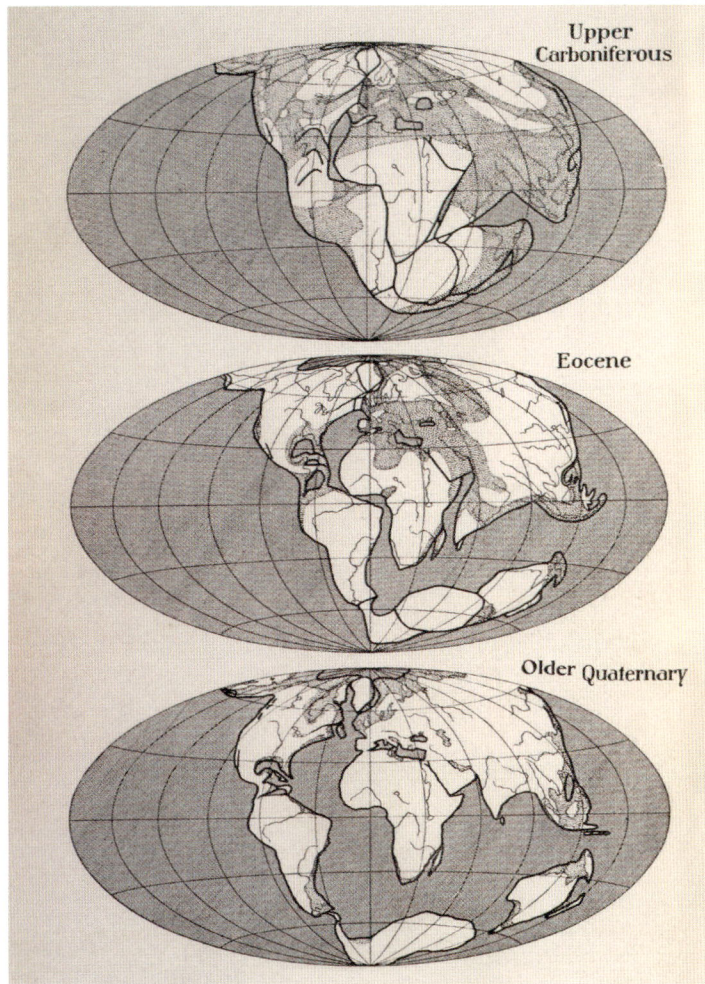

Alfred Wegener (1880–1930), "Reconstruction of the Map of the World for Three Periods According to the Displacement Theory," from *Die Entstehung der Kontinente und Ozeane (The Origin of Continents and Oceans)*, 1824.

Wegener also suggested that many other geological features could be explained by his theory. He proposed that mountains such as the Andes rise up at the crumpling margins of moving continents, or where continents collide—the Himalayas, for example. He also speculated that some oceans are continually widening at the mid-ocean ridges that run along their centers, where fresh new seabed is formed from molten material rising from the depths.

However, Wegener's attempts to explain the causes of continental drift were only partially successful. He correctly suggested that continents and ocean floors float on top of dense, viscous, molten material—a mechanism now called "isostasy." However, Wegener's main error was to propose that the energy for his continental "displacement" derived from the speed at which the Earth rotates, or changes in the axis about which it rotates. These potential energy sources were soon shown to be inadequate, so for fifty more years continental drift remained controversial—a phenomenon without a plausible mechanism to explain it.

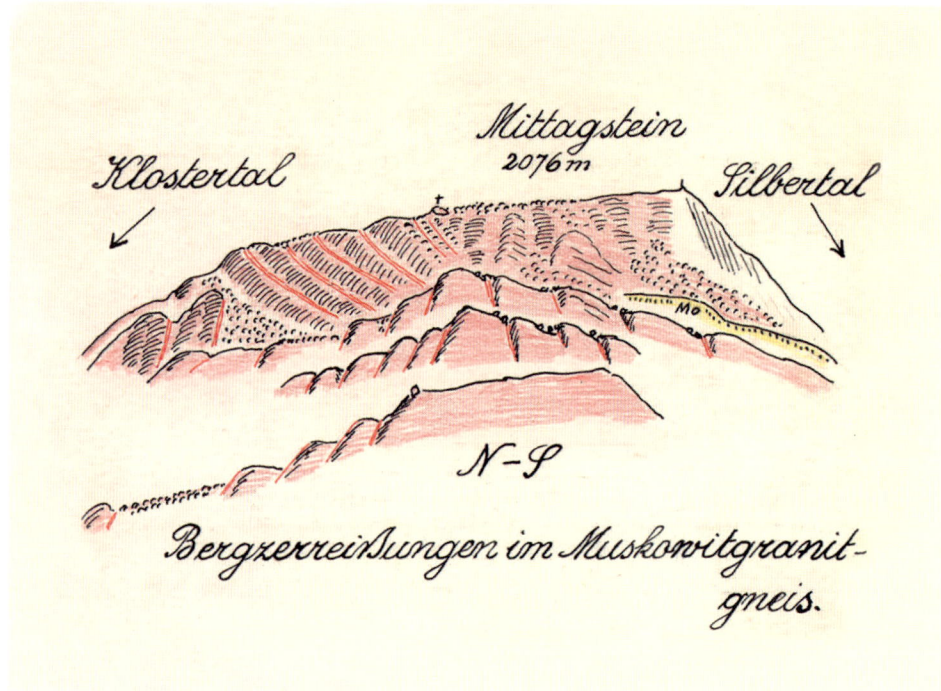

Otto Ampferer (1875–1947), illustration showing a mountain splitting, *c*. 1939.

The Austrian geologist and mountaineer Otto Ampferer was a crucial link between nineteenth- and twentieth-century geology. Working at the University of Innsbruck in Austria, surrounded by the Alps over which he clambered, Ampferer studied the energies that drive mountain building or "orogeny." He proposed that mountains could be formed by the "overthrusting" of one region of crust over another, that mountains could be split asunder, and also that one region of crust could dive under another, a process we now call "subduction." Importantly, he argued that the crust does not itself contain sufficient energy to drive these upheavals and that they must instead be caused by undercurrents, *Unterströmung*, in the molten layers below—setting the geological stage for a century of continental drift and plate tectonics.

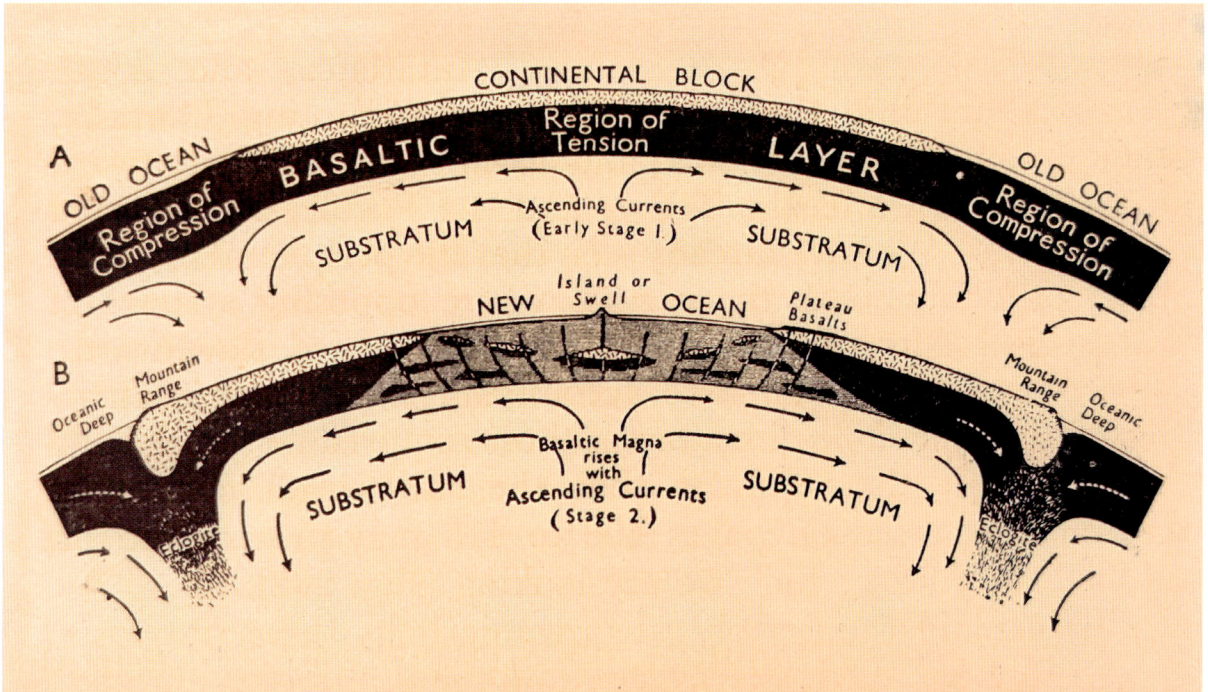

Arthur Holmes (1890–1965), "Diagrams to Illustrate a Purely Hypothetical Mechanism for 'Engineering' Continental Drift," from *Principles of Physical Geology*, 1944.

The English geologist Arthur Holmes (page 45) was a staunch supporter of Wegener's ideas but realized that to gain acceptance, continental drift needed a convincing source of energy to drive it. In a hypothesis similar to Ampferer's *Unterströmung* (see opposite), he suggested that the moving continents ride upon vast convection cells in the mantle, with upward flow splitting oceans apart at their centers to widen them, and downward flow drawing crust deep into the Earth.

Many scientists still believed the mantle to be solid, so Holmes's ideas did not receive universal acceptance, although they were later seminal in the development of the theory of plate tectonics. Indeed, mantle convection remains controversial today—the layers of the mantle and the arrangement of its rolling convection currents are uncertain. There is also no agreement about the speed at which convection occurs—and some argue that much of the viscous mantle may only have circulated once since it formed.

Where does the Earth's energy come from?

LIVING ON AN ENERGY BUDGET

Energy is defined rather obliquely as the ability to make something happen—or, as a physicist would say, it is the ability of a "system" to do "work." Despite this, energy can be measured accurately, usually in *joules*—a unit which can be manipulated alongside standard units such as meters, watts, and seconds.

Under normal circumstances energy cannot be created or destroyed, although it can be converted from one form to another. Indeed, it could be argued that this interconversion of different forms of energy is pretty much what geology is.

First of all, objects can possess *kinetic* energy simply because they are moving—after all, a speeding bullet can certainly make a lot happen. A special type of kinetic energy is *heat* energy, which is the energy a substance contains because of the movement of its constituent particles—heat drives steam and internal combustion engines. The heat in those engines is produced by the release of *chemical* energy in the interatomic bonds within coal or petroleum. Similarly, nuclear power stations derive useful heat

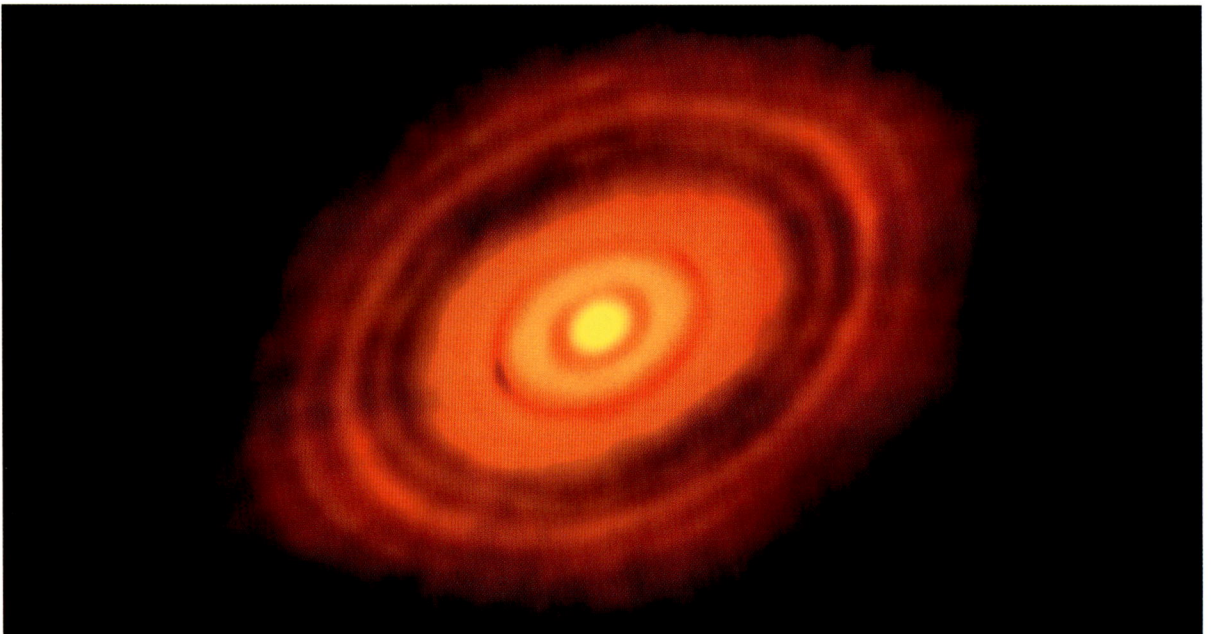

from *nuclear* energy released when the central nuclei of radioactive atoms decay. There is also *elastic* energy, stored in the distortion of a substance—the sudden slippage of the San Andreas Fault in 1906 (page 77) released vast amounts of elastic energy.

There is also *gravitational* energy. Objects can possess energy simply because they are able to "fall." For example, hydroelectric power stations produce electricity from the energy released by water as it falls. Also, a spacecraft falling to Earth gains immense speed—copious kinetic energy which must be dissipated as heat during re-entry. Gravitational energy may have provided the Earth with much of its start-up energy budget. In 1943 the Russian astronomer Otto Schmidt (1891–1956) proposed a model of how the planets of the solar system formed— from a rotating cloud of gas and dust whose particles stuck together to form ever larger particles. Once large enough, these then "fell together" under their mutual gravitational attraction, a process now called "accretion." Remarkably, images of "protoplanetary discs" have now been detected by radio telescopes (see below left) at the stage before they accrete into planets.

These source materials for the Earth *became* hot when vast amounts of gravitational energy were released as they fell together. This heat melted the proto-Earth, allowing dense materials such as iron to sink through the lighter materials and establish Earth's internally layered arrangement—a process called "differentiation."

This is where perhaps half of the planet's internal energy comes from—its initial "falling together." Much of the rest comes from the decay of radioactive elements. So Earth's vast energy store dates almost entirely from when it collapsed into being.

LEFT **ALMA image of the protoplanetary disc around HL Tauri, 2014.**

This image shows the formation of another solar system. HL Tauri is 28 million times further away than our own sun, yet here we can see the disk of material swirling around it, and dark gaps in which planets may be forming.

RIGHT **Fabio Crameri, based on data from J. Huw Davies, "Global Map of Solid Earth Surface Heat Flow," from** *Geochemistry, Geophysics, Geosystems,* **vol. 14, pp. 4608–22, 2013.**

Heat energy does not escape evenly from the Earth's surface—mid-ocean ridges emit much more heat than other regions.

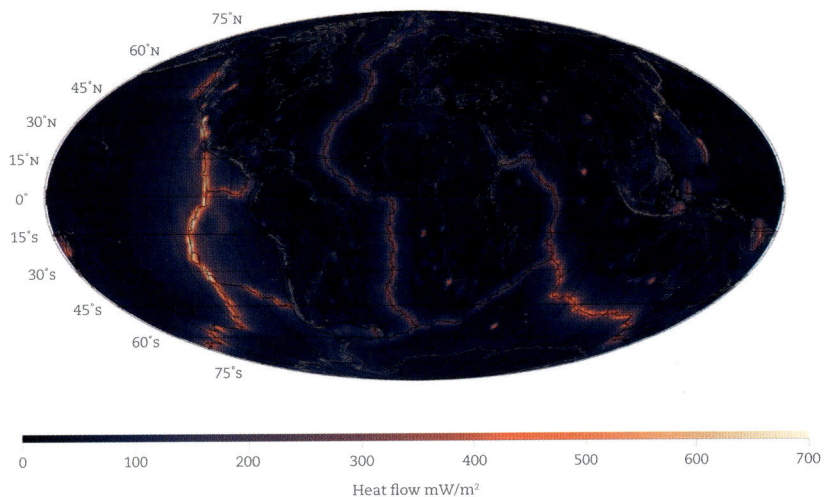

Heat flow mW/m²

**Ott Hilgenberg (1896–1976), cover image for *Vom wachsenden Erdball*
(The Expanding Earth), 1933.**

We now take for granted that continental drift is real, and that it has taken
place across the surface of a planet which has remained essentially the same
size for a long period of time (although its cooling might have caused it to shrink
a little—see page 93). However, we have not always been so certain, and nowhere
is this more apparent than the theory of the expanding Earth.

The Hungarian geologist László Egyed (1914–1970) argued that the reason the
continents' coastlines fit together is altogether different. He suggested that
the globe was once much smaller—so much smaller in fact that the continents
nestled together, covering its entire diminutive surface. Then, as the rocks inside
the planet changed their internal configuration, they expanded, increasing its
volume and stretching its surface area like an inflating balloon. The continents
thus had space to split apart and oceans to form between them. According to
this model, no subduction need occur, because all new oceanic crust formation
is "accounted for" by the swelling of the Earth, rather than requiring crust to be
destroyed elsewhere.

The expanding Earth is certainly a bold counter-theory to continental drift, but
unfortunately there is no evidence for any of its proposed causes or effects. Some
have even claimed the expansion is not due to alterations in the planet's internal
minerals, but progressive changes in the nature of space and time themselves—
although once again, there is no evidence for this.

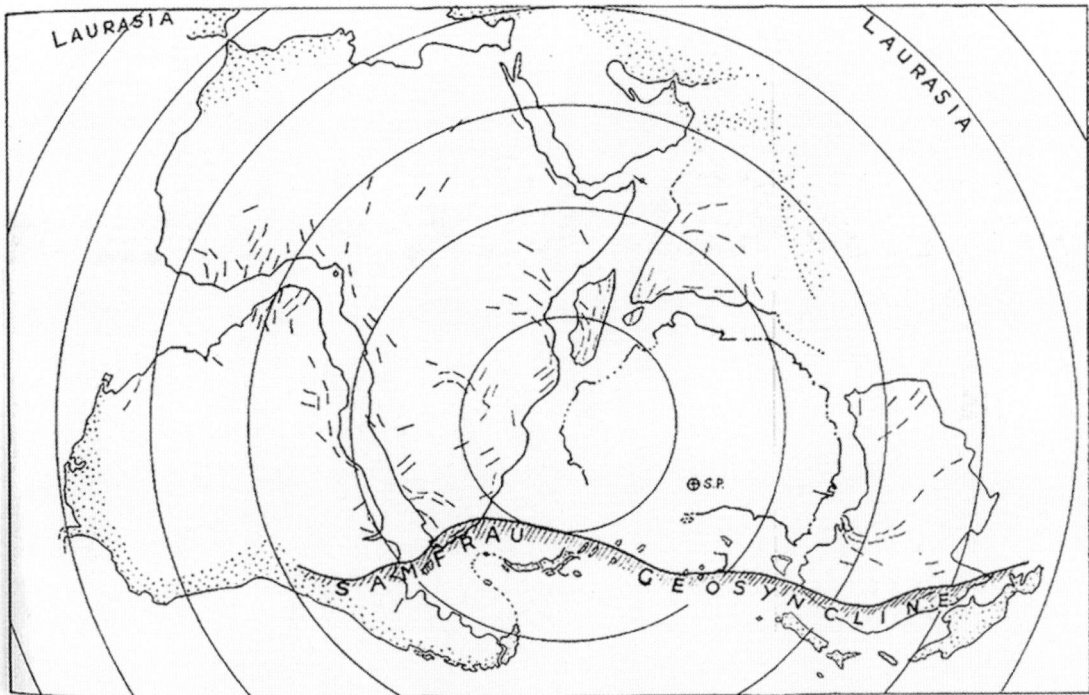

Alexander du Toit (1878–1948), reassembly of Gondwana during the Paleozoic era, from *Our Wandering Continents*, 1937.

Even before continental drift was generally accepted by the geological community, scientists were using it to plot the march of landmasses across the globe. The South African geologist Alexander du Toit was an early supporter of Wegener's ideas and used them to support his theories regarding the origins and history of his native Southern Africa. In particular he suggested there was a time in the past when there were two giant "supercontinents"—Gondwanaland in the south, including what is now Africa, South America, Australia, Antarctica, India, and Arabia, and Laurasia in the north, comprising North America and various pieces of Eurasia.

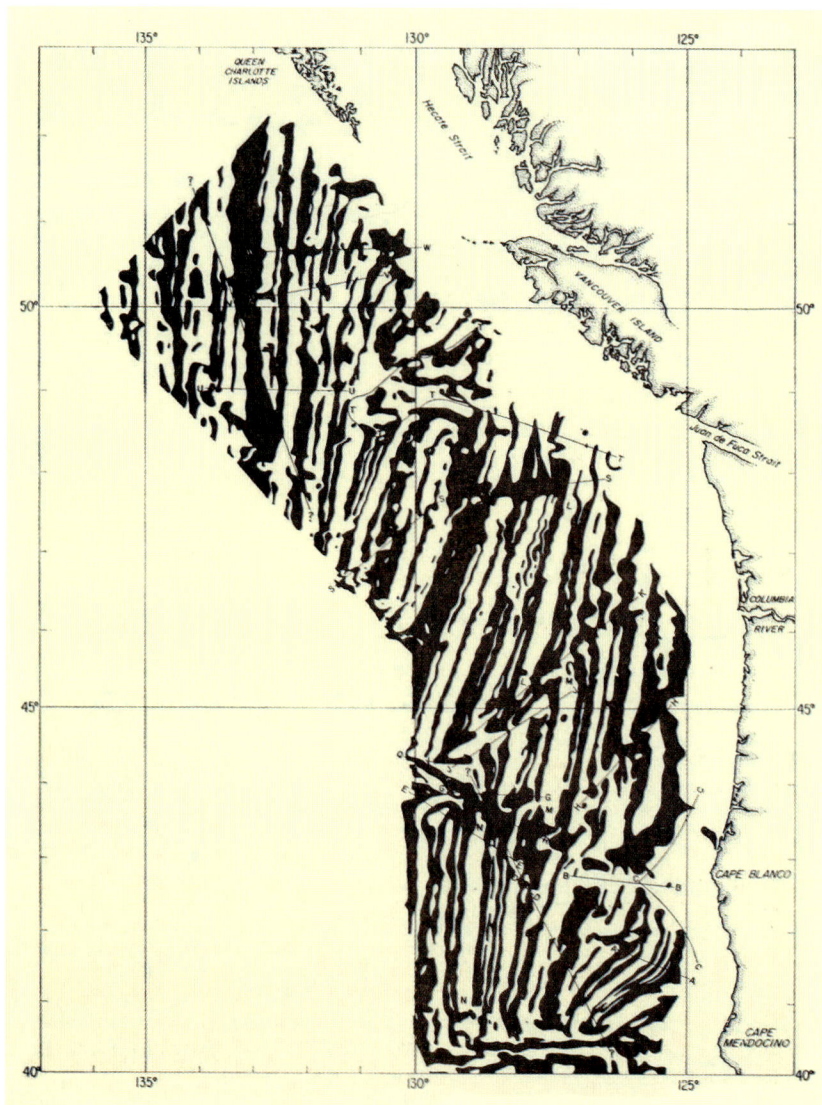

Frederick Vine, summary diagram of total magnetic-field anomalies southwest of Vancouver Island, from "Spreading of the Ocean Floor: New Evidence," *Science,* **vol. 154, pp. 1405–51, 1966.**

It had been known since the late 1920s that the Earth's magnetic field undergoes occasional unexplained and unpredictable reversals, and that igneous rocks retain an imprint of the prevailing magnetic polarity from the time they formed. However, this realization did not feed into the study of continental drift until the early 1960s, when Lawrence Morley (1920–2013) of the Geological Survey of Canada discovered that if ocean crust spreads outward

as new crust is formed at the mid-ocean ridges, then today's seafloor rocks might show regular patterns of alternating north–south magnetization, arranged symmetrically on either side of the ridges.

In 1963 Morley tried unsuccessfully to get his hypothesis published, but later that year striking evidence was published by two British geologists Frederick Vine (1939–2024) and Drummond Matthews (1931–1997) who had clearly had the same idea. Oceanic crust does indeed exhibit beautifully symmetrical "stripes" of magnetization on both sides of the ridge—the first direct evidence of ocean widening, and a central element of the theory of plate tectonics.

Marie Tharp (1920–2006), American geologist and oceanographer, constructing a map of the Atlantic Ocean floor in the early 1950s.

The floor of the deep sea is notoriously difficult to study, although from the middle of the eighteenth century geologists had suspected the presence of an elevated mid-ocean ridge running along the center of major oceans. It had even been suggested that it might play a role in continental drift.

In the 1950s, the American cartographer Marie Tharp elucidated the topography of the deep oceans with a clarity which was to prove crucial to the later development of plate tectonics. Working partly at the Lamont Geological Observatory at Columbia University, Tharp used bathymetry—depth soundings gathered by research ships which she, as a woman, was not permitted to board—to create accurate coast-to-coast topographic "sections" of entire oceans. She revealed the striking prominence and extent of the Atlantic mid-ocean ridge, and crucially its central rift valley where the ocean is now known to "split apart" (page 92). The rift valley, and its seismic activity, were important new pieces in the conceptual jigsaw that would soon become plate tectonics.

John Tuzo Wilson (1908–1993), illustration from "Evidence From Ocean Islands Suggesting Movement in the Earth," *Philosophical Transactions of the Royal Society of London Series A,* vol. 258, pp. 145–67, 1965.

The central theory of modern geology is plate tectonics, formulated in the mid-1960s as many different lines of evidence finally converged. With hindsight it seems inevitable that plate tectonics was the great geological solution, but the varied nature of the evidence supporting it meant the theory did not leap from the mind of one person, but rather from communal discussion between geologists from around the world.

Plate tectonics is based upon the idea that the Earth's crust and uppermost mantle (which together constitute the "lithosphere") is divided into many raft-like plates which float on a denser but more fluid layer of the mantle (the "asthenosphere"). There are two types of plate—continental crust, which is rich in aluminum and sodium and is often billions of years old, and oceanic crust, which is rich in iron and magnesium and tends to be only tens of millions of years old because it is continually produced and destroyed. Continental crust is thicker but lighter and so "floats high" on the mantle, whereas oceanic crust is denser and thinner and "sits lower," which is why it is usually covered by water.

Some oceans such as the Atlantic are currently widening, as new igneous crust is produced at their seismically active mid-ocean ridges. Most of the rest of the world's seismic activity takes place where tectonic plates meet at "boundaries"—here plates may collide and crumple, or slide over, under, or along each other. Indeed, it is the worldwide network of plate boundaries which largely explains the lines of seismic activity in Robert Mallet's 1857 seismographic map (page 73).

Where continental plates collide and crumple, mountains may be formed, and a good example of this is the Himalayas, a "young" mountain range where the Indian and Eurasian plates are colliding at a speed of 67 mm (2.5 in.) per year. In contrast, where continental and oceanic plates are moving toward each other, the denser oceanic plate often dives or "subducts" under the continental—many earthquakes occur in subduction zones, and as the diving crust is obliterated, rising molten material drives volcanic activity. Subduction also often occurs when two oceanic plates come together, and this downward movement creates the world's deep-sea trenches and most volcanic island arcs. Finally, where plates slide horizontally "along" each other, such as at the San Andreas fault, the main result is usually earthquakes, with little sign of volcanism, or destruction or creation of new crust.

These maps are from one of the series of scientific papers around which the theory of plate tectonics coalesced. Here it is proposed that underlying convection drives the movement of tectonic plates away from mid-ocean ridges (dashed lines) and toward collision boundaries (solid lines) where mountain ranges are formed.

Plate tectonics is an elegant theory—various lines of evidence support it, and it explains many of the planet's geological phenomena. However, it is not perfect. In particular, there is no agreement about how the Earth's internal energy drives plate movements. It was once thought that convection currents in the mantle similar to those suggested by Arthur Holmes (page 83) wafted plates across the surface of the globe. However, while the mantle certainly does play a role in tectonics, many geologists believe that the asthenosphere is too weak to "drag" large plates of lithosphere around on its own. Instead, it is possible that creation of new crust and subduction of old crust may "push" or "pull" tectonic plates around.

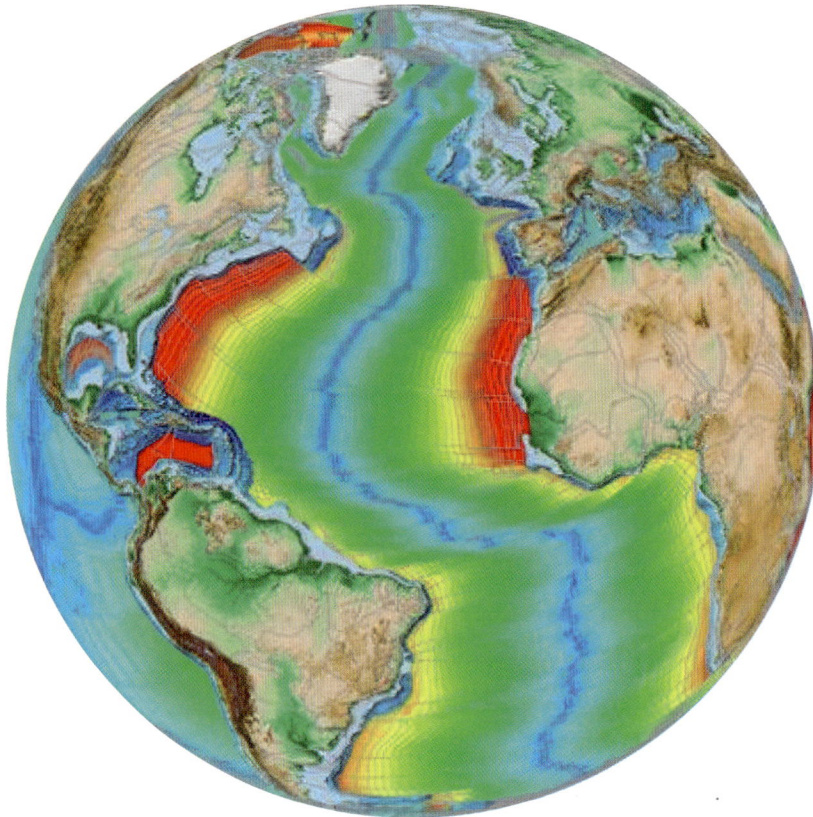

R. Wilson, et al., illustration from "Fifty Years of the Wilson Cycle Concept in Plate Tectonics: An Overview," *Geological Society, London, Special Publications*, vol. 470, pp. 1–17, 2019.

No sooner had plate tectonics been proposed, than it was suggested to occur *cyclically*. In 1966 one of the theory's originators, John Tuzo Wilson, published a paper with the intriguing title "Did the Atlantic Close and then Re-Open?" He argued that not only do now-distant landmasses support closely related fauna, but conversely, now-adjacent regions appear to have been inhabited by distantly related creatures. He proposed that, although the continents are now moving apart following the fragmentation of Pangaea, they previously moved together to form it.

Tuzo Wilson's idea has evolved into what is now called the "supercontinent cycle." We now believe that Pangaea, which existed around 200 million years ago, was formed from the fragments of an earlier supercontinent, Pannotia, which itself coalesced 600 million years ago. And Pannotia was formed from the fragments of Rodinia, which contained most of the Earth's landmass 900 million years ago. And Rodinia was preceded by a series of progressively more mysterious "ancestor" supercontinents.

Obviously, the supercontinent cycle is about as enormous as a geological theory can be, but it raises enormous questions as well. Why do continents undergo this cycle of alternating consolidation and separation? When did the cycle start and why? Why do supercontinents often break apart near the same fracture lines as their predecessors? And where did all the continental crust come from in the first place?

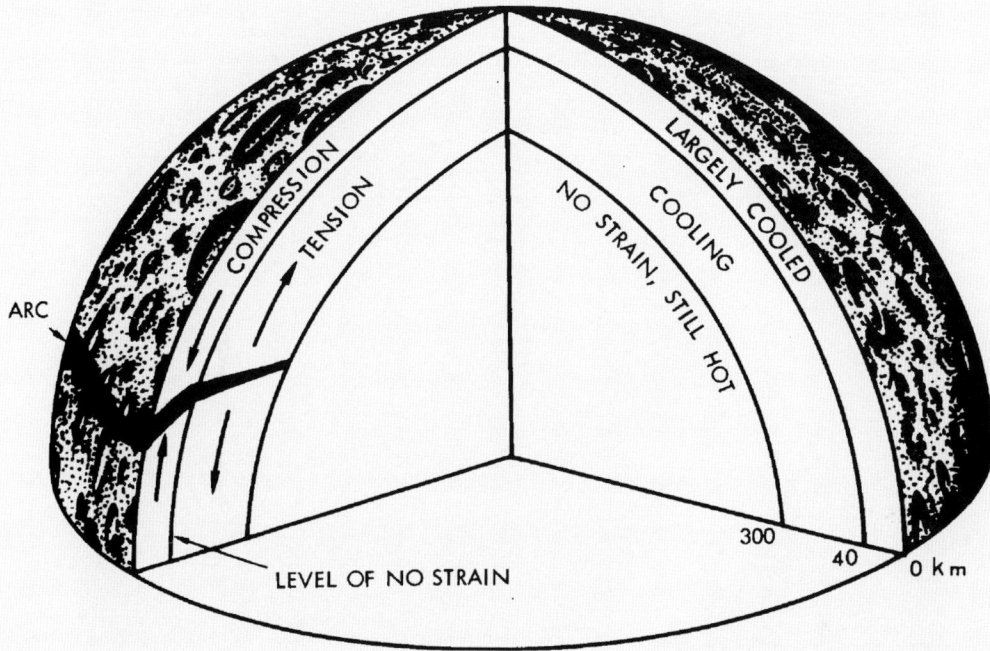

Han-Shuo Liu, states of thermal strain of Mercury, from "Thermal Contraction of Mercury," NASA *Technical Note* D-5867, 1970.

Although plate tectonics had finally demolished the "expanding Earth" theory as a cause of continental drift (page 86), there remained the possibility that some of the planet's features might be explained by it *shrinking* as it cooled. This is a reasonable suggestion, since most substances do indeed contract as they cool, yet it seems unlikely that the effects of any tiny shrinkage would be visible among the plate wanderings, collisions, and subductions of plate tectonics.

However, the shrinking planet theory does in fact seem to apply elsewhere—to the solar system's smallest planet, Mercury. Although small planets are assumed to lose their internal heat faster, Mercury shows evidence of retaining some of it—unlike the larger planets Mars and Venus, it still drives its own magnetic field, and some think tectonic activity may have ceased there only recently. Nonetheless, it is sometimes described as a "one-plate planet" and its surface is relatively quiescent now. It does, however, exhibit definite signs of cooling shrinkage—elevated ridges called "lobate fault scarps," not unlike the wrinkles that appear on fruit as it shrivels. A similar process may also have occurred on our own Moon.

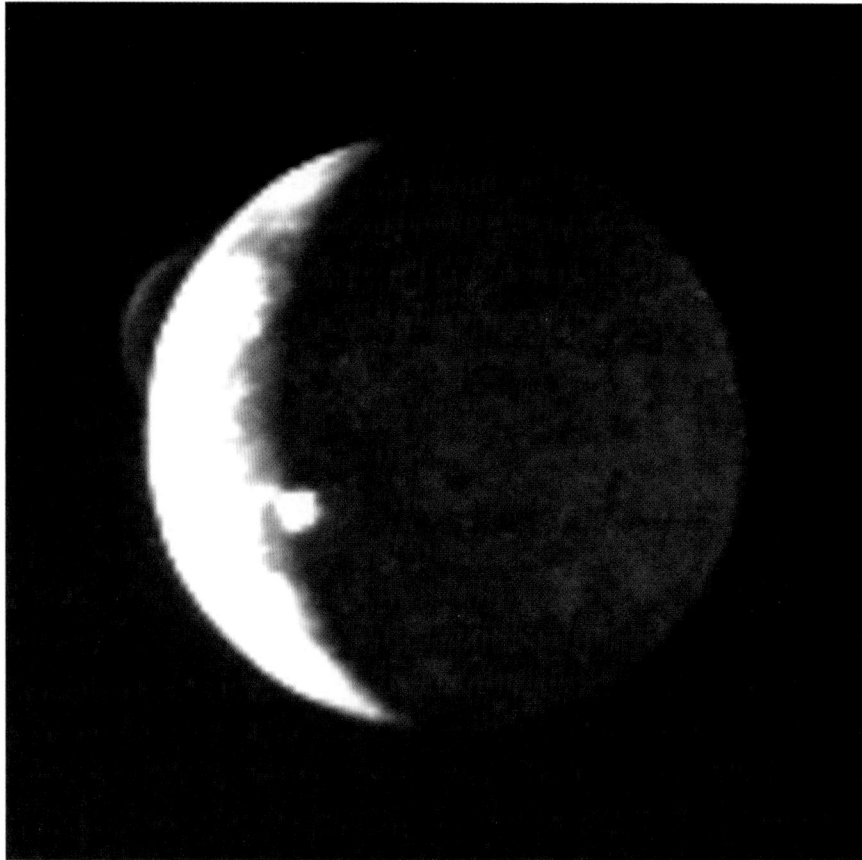

NASA, Voyager 1 discovery photograph of a volcanic plume on the Jovian moon Io, 1979.

Ever since Alfred Wegener erroneously proposed that the Earth's rotation drives continental drift (page 81), geologists and astronomers have wondered whether celestial orbits and rotations can affect geological processes, even on a small scale. In fact, just as the gravitational effects of the Sun and Moon cause the Earth's seas to rise and fall, they also pull its solid crust up and down. The Sun, Moon, and Earth interact to cause a complex variety of tidal effects, and Earth's surface may rise and fall by as much as half a meter (almost two feet).

If tidal effects have any effect on processes such as earthquakes, it is probably subtle. However, evidence of far stronger tidal effects has been discovered on Io, the third-largest moon of Jupiter. Jupiter is extremely massive, and its gravity distorts Io into an egg-shape with its point facing the

giant planet. In addition, Io's orbit is elliptical, not circular, so the degree of "pointiness" of the egg shape is continually changing. The resulting repeated distortion of Io imparts large amounts of heat energy, and this is why Io turns out to be the most geologically active body in the solar system— and probably has been for the last 4,500 million years.

Although there are more detailed and colorful images of Io, this picture is significant as it represents the first time an active volcano was discovered on another world—the diaphanous curve at the ten-o'clock position is a volcanic plume, imaged by the Voyager 1 spacecraft on March 8, 1979, and identified, to her great surprise, by Linda Morabito (1953–) of the Jet Propulsion Laboratory in Pasadena, California.

François Gauthier-Lafaye, location of the Oklo natural nuclear reactor, 1972.

A major source of the Earth's internal energy is the "decay heat" of unstable radioactive isotopes, and a surprisingly large fraction of this energy is liberated within the Earth's thin crust (page 61). To the scientifically inquisitive, this raises the question of whether sufficient radioactive material has ever been naturally concentrated in one place to initiate a nuclear fission chain reaction. And perhaps surprisingly, the answer seems to be yes.

In 1972, French prospectors discovered unusual deposits of uranium in the Oklo region of the Haut-Ogooué province of Gabon, Central Africa. They noticed that the isotope uranium-235 made up only 0.717 percent of the sample, rather than the more usual 0.720 percent. This may not sound like much, but it is similar to the uranium-235 depletion seen in artificial nuclear reactors, and the subsequent detection of telltale fission products at Oklo confirmed that natural fission had indeed once occurred deep in the African crust. Thus, perhaps 1,800 million years ago it seems the Oklo region accumulated sufficient nuclear fuel, and enough water to "moderate" the chain reaction, for the world's first nuclear reactor to "switch on."

Air as liquid, water as rock, sand as mist, metal as rain

GEOLOGY IN OTHER WORLDS

In recent years geologists have realized that Earth is one geologically active body among many. Within the solar system there are worlds more active than ours, such as Io, and worlds less active, such as Mercury. It is also clear that geological processes occur across a far wider range of temperatures than exist on Earth, and while some of these may seem superficially familiar, the chemistry and physics behind them can be utterly unearthly.

The geysers of Triton sit at the cold end of the scale. Triton is the largest moon of Neptune and has a surface temperature of -240°C (-400°F)—so when the Voyager 2 probe passed by in 1989 not much was expected from a world so remote from the Sun's warmth. However, the surface of Triton has turned out to be surprisingly varied, and seems to be regularly "resurfaced" by "geological" processes.

Although Triton is so cold that water would exist as a hard, brittle mineral, its surface is rich in carbon monoxide, nitrogen—which in gaseous form constitutes most of Earth's air—and the simplest hydrocarbon, methane. Although far away, the Sun's energy seems to

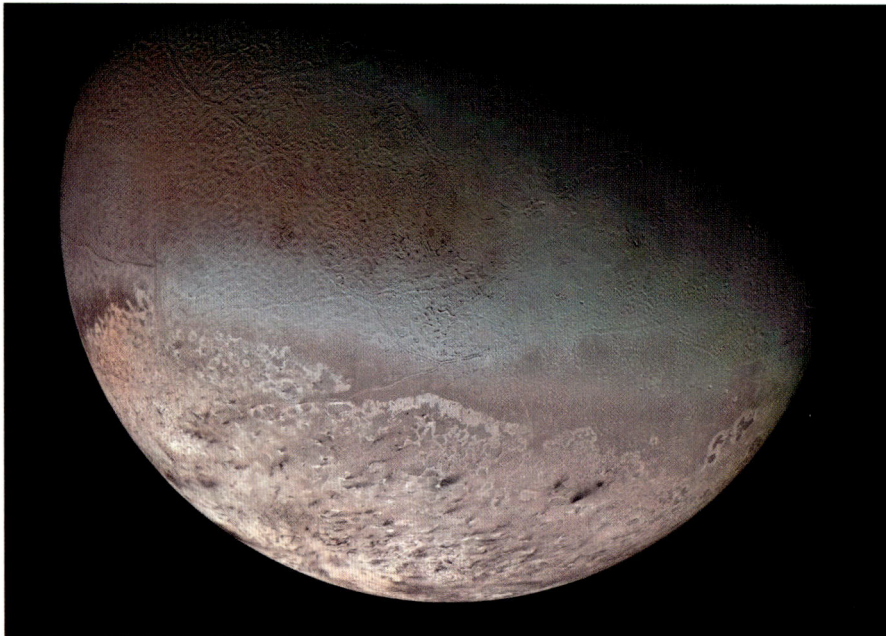

NASA, Voyager 2 global color mosaic of the Neptunian moon Triton, with dark marks from geyser plumes, 1989.

have been strong enough to combine methane molecules into a variety of complex products that give the moon a delicate pastel coloring.

Most surprising of all was the discovery of nitrogen geysers blasting plumes of dark material 8 km (5 mi.) above the surface, which then drift slowly westward forming dark "smears." Unexpectedly spotted by Voyager team member Larry Soderblom (1944–), it is thought solar energy causes this phenomenon. As Voyager 2 careened past Triton, most geysers sat at latitudes which directly faced the Sun—and it is thought the icy, translucent surface traps incoming solar energy, which warms subsurface nitrogen to explosion point, in a greenhouse process analogous to how methane and carbon dioxide trap energy in Earth's atmosphere.

In stark contrast, one of the hottest worlds ever discovered is WASP-76b, orbiting a star 6,000 million million kilometers away. This exoplanet orbits its star at one thirtieth of the distance that Earth orbits the Sun, and its surface is heated to around 2,200°C (4,000°F). The planet has an incandescent glow, and many of the chemicals we think of as rocks or metals on Earth exist as liquids or even vapors. Various studies have suggested that WASP-76b's atmosphere contains quartz (silicon dioxide), alumina (aluminum oxide), and metallic iron.

Like Earth's orbiting Moon, WASP-76b's rotational kinetic energy was once slowly expended in distorting and heating it, so that it now rotates at a much slower speed. As a result, the exoplanet is now "tidally locked" so that the same "hot" hemisphere always faces its star. And although the data is disputed, it has been suggested that metallic iron vapor formed on the sunward side is continually windblown around to the "less hot" side, where it condenses and falls as metal rain.

Heather Knutson, NASA, A map of the day-night contrast of the extrasolar planet HD 189733b, another extrsolar planet, 2007.

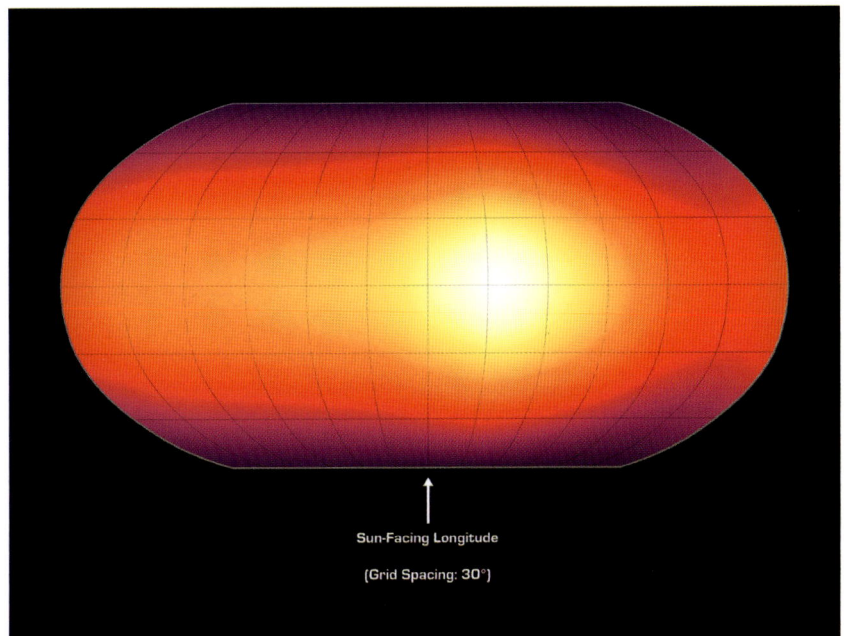

Sun-Facing Longitude

(Grid Spacing: 30°)

Mount Erebus in Antarctica, NASA Earth Observatory image by Lauren Dauphin using Landsat data from the U.S. Geological Survey, 2023.

There are fewer than ten lava lakes known on Earth. Here, energy reaches the surface at a relatively constant rate—the energy flow is insufficient for a volcano to undergo major eruptions, but it is not so low that the molten rock solidifies. The result is a "lake" of lava exposed to the atmosphere for years at a time, often stirred by the venting of hot gases, and visible from space as a glowing red blob in this image. Discovered in 1972, one of the most dramatic is that on Mount Erebus in Antarctica, the most southerly active volcano in the world. At seventy-seven degrees south, Erebus is permanently snow-covered, but aerial and satellite images show a continually roiling lava lake among the frigid surroundings.

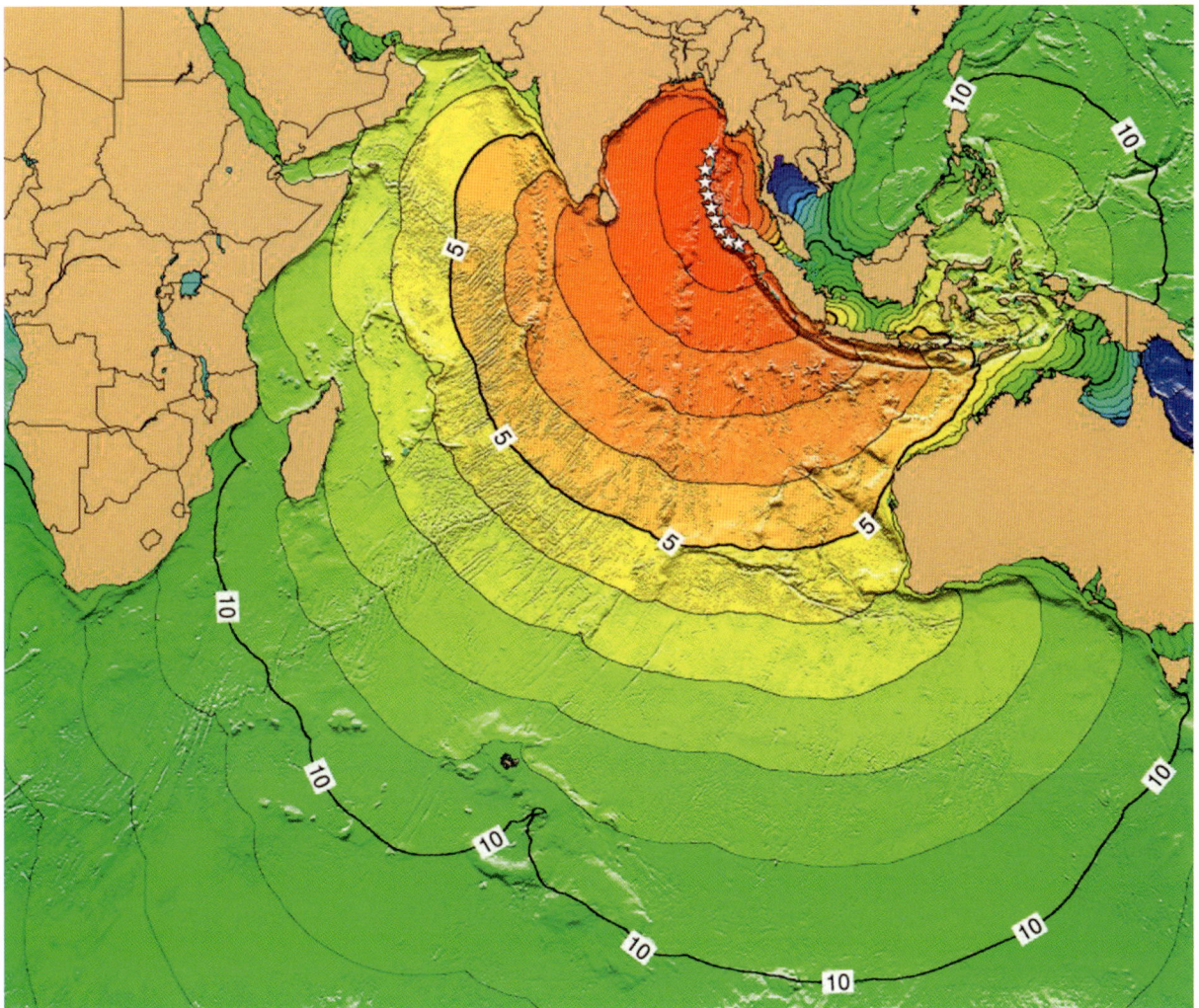

National Oceanic and Atmospheric Administration, tsunami travel time map for the 2004 Indian Ocean tsunami.

On the morning of December 26, 2004, an earthquake off the coast of Sumatra released around a million million million joules of energy into the surrounding crust and ocean, leading to one of the worst natural disasters in history. Displacement of the crust caused a tsunami that propagated across the Indian Ocean over the coming hours, inflicting destruction on coastal communities as far away as East Africa—the map shows the time taken, in hours, for the tsunami to travel from its source (the line of stars) to progressively more distant regions. In many places the sea initially receded from the coastline, after which a barrage of waves of up to 9 m (30 ft) high inundated the land—killing people and animals, destroying buildings and infrastructure, and contaminating croplands with salt. The damage was made worse by the energy and configuration of the quake, but also the relative unpreparedness of nations around the Indian Ocean. Readiness for tsunamis was much greater in regions facing the Pacific—after all, "tsunami" is a Japanese word meaning "harbor wave."

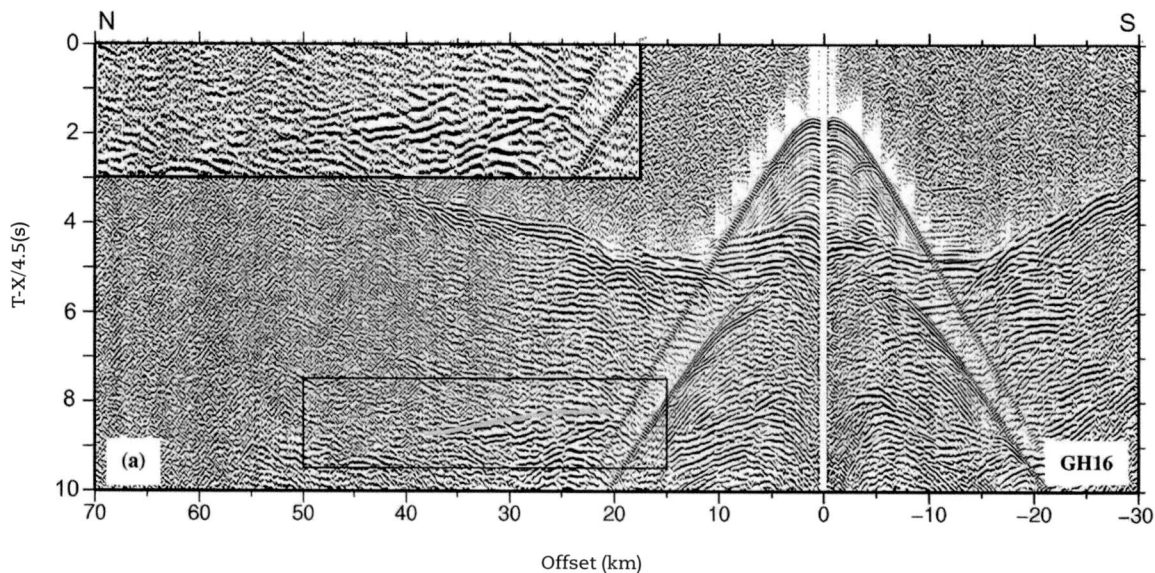

The seismic cross-section figure with axes labeled "N" (top left), "S" (top right), "T-X/4.5(s)" on the vertical axis (0 to 10), and "Offset (km)" on the horizontal axis (70 to −30). Labels "(a)" at lower left and "GH16" at lower right.

B. Bouyahiaoui, et al., representative section recorded within the deep basin along the Annaba profile, illustration from "Crustal Structure of the Eastern Algerian Continental Margin and Adjacent Deep Basin: Implications for Late Cenozoic Geodynamic Evolution of the Western Mediterranean," *Geophysical Journal international*, vol. 201, pp. 1912–38, 2015.

The deep inner workings of our planet can now be investigated as never before: this image shows the kinds of large and complex seismological data set which may be analyzed by modern computing power, including in this case a seismological cross-section though a subduction zone where the edge of an oceanic plate is diving under a continental plate. Subduction causes earthquakes and also generates volcanoes, possibly because it changes the melting properties of the mantle, or even because it pulls seawater down into the subduction zone. Subduction may also actually drive tectonics to some extent, "dragging" plates across the globe. (The pale blue line indicates the data selected for use in assessing seismic wave speeds.)

Tectonic processes such as subduction power our planet's geology, and we almost take them for granted. However, ongoing tectonic activity has not been detected elsewhere in the solar system, hinting that the Earth is perhaps somewhat "special" in this respect. Also, many geophysicists think that in the Earth's distant past, perhaps over 2,500 million years ago, the mantle may have been too hot to support plate tectonics as we now know it. However, this is far from settled: others set the date for the onset of "modern" tectonics soon after the planet's formation, or as recently as 800 million years ago. And if, as is entirely possible, the oldest continental crust pre-dates the onset of tectonic activity, then we will need entirely new theories to explain it.

The Himalayas from the south, photographed by astronaut Don Pettit, ISS Crew Earth Observations Facility and the Earth Science and Remote Sensing Unit at Johnson Space Center, 2012.

Apart from the mid-ocean ridges, the most striking topographical feature on Earth is the Tibetan Plateau. Covering an area more than ten times the size of the United Kingdom, including vast areas of China and six bordering countries, much of it lies around 5,000 m (16,400 ft) above sea level. Considering Mount Everest is "only" 8,848 m (29,028 ft) high, it is clear that nowhere on Earth is such an enormous area as elevated as the Plateau of Tibet—for example, it has profound effects on the world's climate because its bulk "pokes through" much of the atmosphere. Formed over the last seventy million years as various tectonic plates collided, the plateau also has effects deep below the surface. Because the dense, solid crust floats on top of the lighter, viscous mantle, very elevated regions of crust must be buoyed up by deep "roots"—a phenomenon called "isostasy," similar to how a tall iceberg has a large submerged portion to buoy it up. And indeed, the crust beneath Tibet is an unparalleled 70 km (40 mi.) thick.

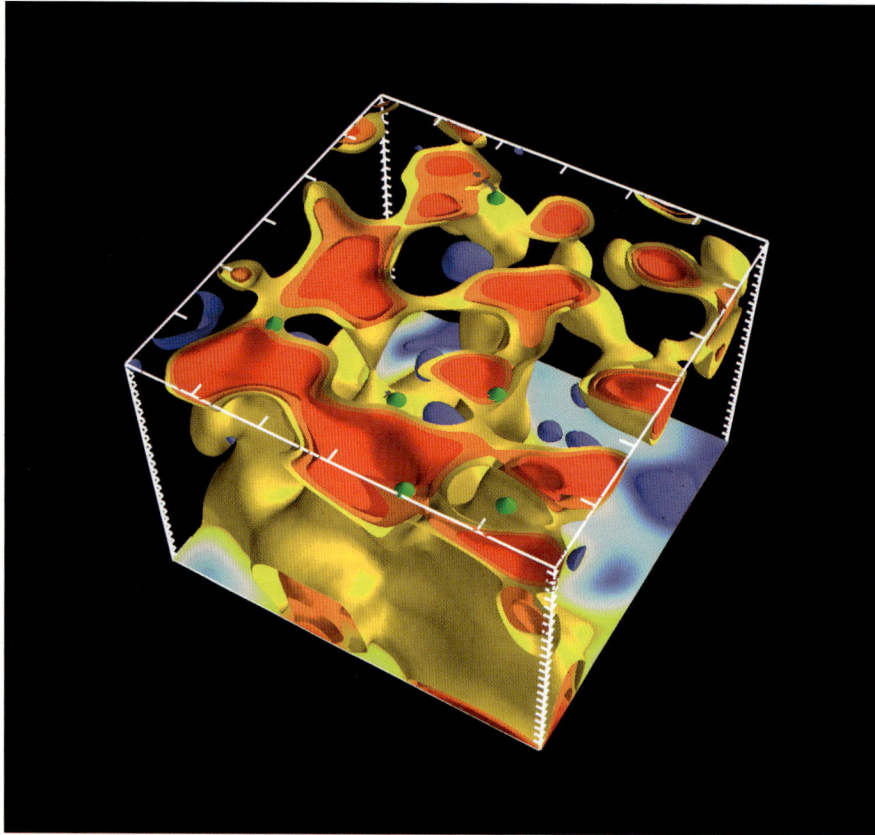

Scott French and Barbara A. Romanowicz, three-dimensional imaging of the Pacific Superswell region, from "Broad Plumes Rooted at the Base of the Earth's Mantle Beneath Major Hotspots," *Nature*, vol. 525, pp. 95–99, 2015.

It has long been suspected that isolated volcanic "hotspots" such as Yellowstone, Hawaii, Samoa, and Iceland occur over narrow rising plumes of hot material in the mantle (see page 75) although the existence and nature of these plumes are controversial. They probably arise from the deep mantle, from regions into which subducted tectonic plates have been sinking for at least the last 300 million years.

This image is from a research study in which seismic records from all over the world were analyzed to generate three-dimensional maps of the deep mantle. The upper surface of the cube represents a quadrilateral region of the Earth's surface corresponding to much of the eastern Pacific Ocean—the small green cones are islands. The mantle underlying this region is represented by the underlying parts of the cube. Subtle variations in the speed of seismic waves in different parts of this cuboidal region of mantle are depicted in shades of red, yellow, and blue. The results showed the presence of plumes originating from the core–mantle boundary and ascending to within 400–1,000 km (240–600 mi.) of the surface. The plumes are near-vertical, and may be long-lived enough to leave "trails of evidence" such as the Hawaiian island chain as the tectonic plates glide above them.

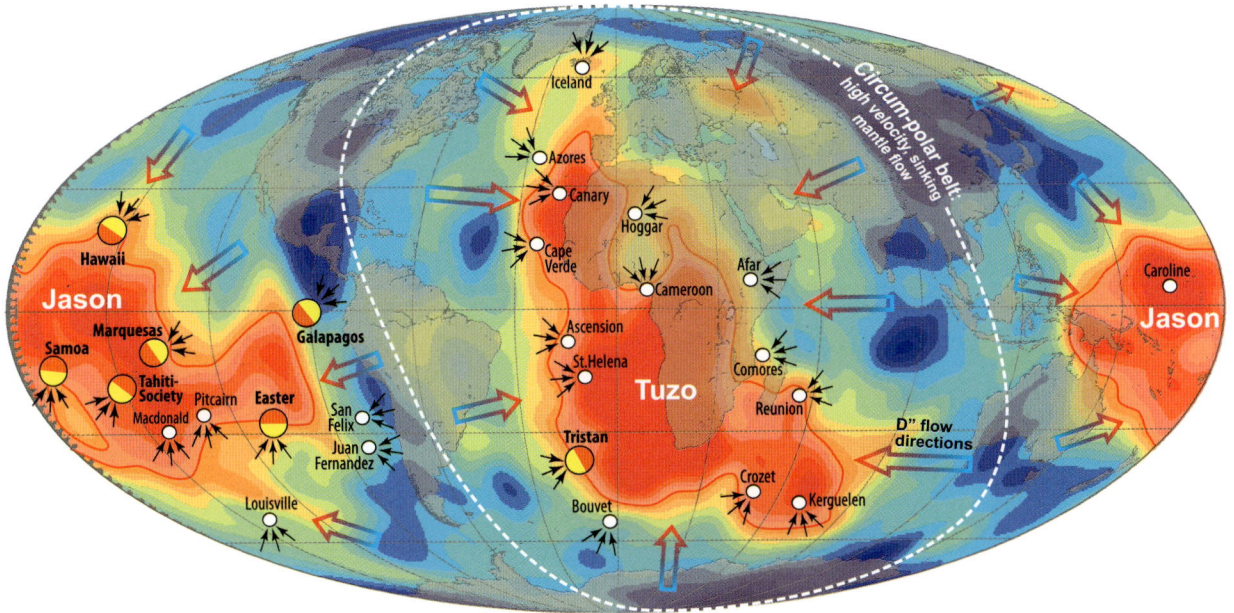

Björn Heyn, et al., tomography model showing the two LLSVPs and the circumpolar belt of subducted material surrounding them, from "Core-Mantle Boundary Topography and its Relation to the Viscosity Structure of the Lowermost Mantle," *Earth and Planetary Sciences Letters*, vol. 543, article 116358, 2020.

It is becoming clear that the deep mantle and its boundary with the core are not unvarying, homogenous regions. Indeed, modern seismological techniques suggest that the deep mantle is characterized by two enormous "Large Low Shear Velocity Provinces" (LLSVPs)—one centered roughly under Angola, and the other under Samoa. However, this map of the velocity of seismic waves deep in the mantle shows that both LLSVPs are extremely extensive (orange) and they are separated by a high velocity seismic region (blue) tracing out a great "circum-core" loop running from under the North Pole, along the line of the Americas to the South Pole, and returning north under Australia, China, and Siberia.

The two LLSVPs have been described as "thermochemical piles": great heaps of mantle material which may differ from surrounding regions in density, viscosity, and chemical makeup. Their movement and weight may even distort the underlying core-mantle boundary to render the core less spherical—in an analogous way to how, "higher up," heavy regions of crust depress the underlying mantle. Thus, this mysterious core-mantle boundary thousands of miles under our feet may have its own great topographic basins, domes, ridges, and trenches.

ABOVE Evidence of glacial rebound in Bathhurst Inlet, western Nunavut, Canada, 2013.

The continental and oceanic tectonic plates float on top of the asthenosphere of the mantle because they are less dense. One cubic metre of mantle has a mass of around 3,300 kg (7,300 lb), while the same amount of oceanic crust is 3,000 kg (6,600 lb) and continental crust is only 2,700 kg (6,000 lb).

Long before the advent of plate tectonics, geologists wondered whether the "isostatic equilibrium" of crust floating on mantle might be disturbed when that crust becomes loaded by heavy glaciers and there is now a great deal of evidence suggesting that this is the case. During ice ages, so much water is sequestered in glaciers on land that global sea levels fall by hundreds of feet. This removes load from the oceanic crust and the continental shelves (the low-lying edges of continental landmasses which are covered by shallow water). Conversely, glaciers add load on top of the continental crust, which sinks as a result.

Once the glaciers recede, the crust "rebounds" to its previous level, so that most ocean crust is currently falling and previously glaciated continental crust in mid to high latitudes is rising.

RIGHT Julien Aubert, European Space Agency, simulation of the magnetic field in Earth's core, 2019.

The Earth's magnetic field is driven by heat energy from its core. Although heat is still being produced inside the Earth by radioactive decay, it is unlikely that much of this happens in the core, so that central iron alloy ball has progressively cooled throughout most of the planet's existence.

The solid inner core (page 79) may even be a single vast crystal—and it is gradually getting larger. As the core cools, the lower layer of the liquid outer core solidifies onto it, enlarging it by perhaps one millimeter a year. This solidification liberates heat energy, and also releases oxygen into the liquid outer core—and both of these may cause the turbulent churning of the outer core. Vast rivers of molten iron writhe in the outer core and it is this which generates the Earth's magnetic field (in the image, the iron currents are yellow and the magnetic field lines are red and blue). This turbulent flow seems to be quite "lumpy," however, and the introduction of new blobs of molten metal into the flows may explain the magnetic "jerks" which occur every few years when the magnetic field subtly but suddenly alters its direction.

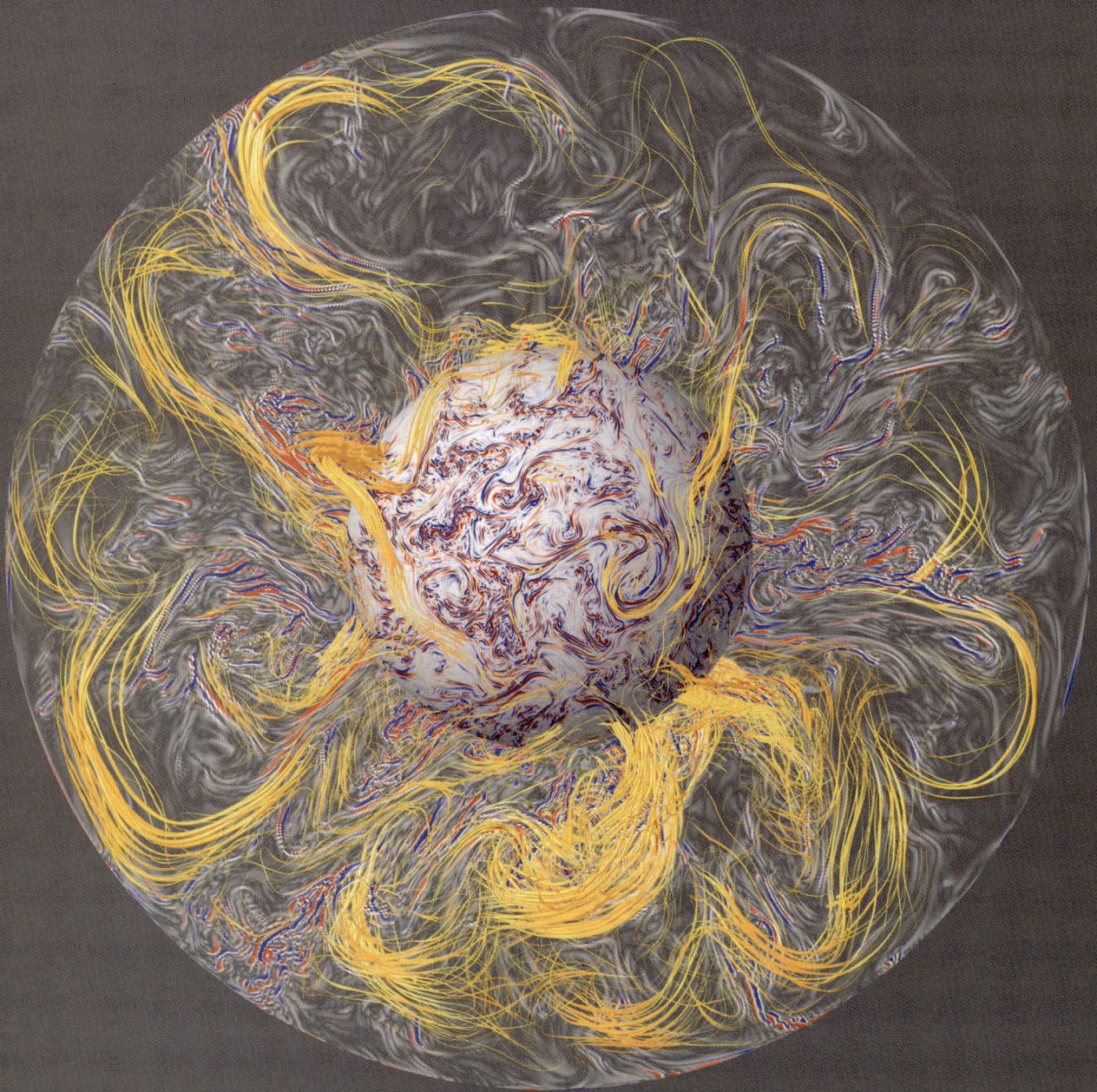

3

PROCESS

Slow deposition of sediments, and energetic ejection of igneous rocks, are not sufficient to explain the geological wonders we see around us. From the tiniest crystals to the largest landforms, the crust beneath our feet is molded by myriad interconnected processes of creation, modification, and destruction.

Polarized light photomicrograph of a thin section of dunite from Lake Balaton, Hungary, 2012.

First achieved in 1829, passing polarized light through thin slices of rock—in this case three hundredths of a millimeter—allows its internal structure to be visualized.

"When the aggregate amount of solid matter transported by rivers in a given number of centuries from a large continent, shall be reduced to arithmetical computation, the result will appear most astounding … "

Charles Lyell, 1833

T he thinkers of Ancient Greece and the Islamic Golden Age set the tone for much of modern geology. They pondered how stones form and how mountains form, and whether rocks change deep within the Earth, and they realized that all may be destroyed and carried away by the action of water. And in the medieval period thinkers formed philosophical frameworks to explain the mechanisms which forge the world around us, and started the difficult process of cataloguing the Earth's constituents and mechanisms.

The Catalan philosopher and knight Ramon Llull (*c.* 1230–1316) was the originator of the "animal, vegetable, or mineral" classification system. He proposed an ascending "scale of being," a *scala naturae* leading from base material up to God—but in his system, *lapis* (stone) occupied a lowly position below "flame," plants, and animals. In contrast, the French cleric-philosopher Jean Buridan (*c.* 1300–1359) took a more mechanistic approach, seeking reasons for why the Earth has two dissimilar regions: land and sea. He argued that the present arrangement is only temporary, and that while water is wearing the land away in wetter parts of the world, it is simultaneously depositing it elsewhere in the dry. Thus, the wet and the dry vie continually, and in so doing seek some sort of equilibrium, or perhaps even a cyclical rhythm.

The study of the origin and classification of minerals has a

Ramon Llull (*c.* 1230–1316), "Scala Naturae" from *Liber de Ascensu et Descensu Intellectus* (Book on the Ascent and Descent of the Mind), c. 1305.

Nicolas Regnesson (d. 1670), portrait of Jean Buridan (c. 1300–1359).

long history and has proved to be one of the hardest geological nuts to crack. It is worth emphasizing that it has been much harder to classify minerals than animals or plants. Even today geologists still study why minerals with similar chemical ingredients are so different. In addition, it is all very well to believe that sediments are deposited over time, but how do they "become" rock? Also, we know that lava comes out of the Earth, but how does it "become" soil?

Although many minerals were known since antiquity, often for practical purposes, Ibn Sina in the eleventh century (page 22) was one of the first to speculate about the formation of stones, and how malleable, brittle, soluble, and "oily" minerals might arise. Later, the French entomologist, René Antoine Ferchault de Réaumur (1683–1757) described how stones are worn away, forming a "lapidifying juice," which then provides the raw material for the creation of new crystals in cavities within existing rocks.

Progress was understandably slow in a scientific world before the periodic table or the atomic theory of matter. Yet fortunately, crystals offered a convenient way around this problem. This is because their visibly regular shape parallels their internal geometric arrangement of atoms—first named *molécules intégrantes* early in the nineteenth century. Steno (page 16) had already noticed that the angles between the faces of a particular crystal type are always the same, and indeed are characteristic of that mineral. However, eighteenth-century attempts at mineral classification were halting, sometimes focusing on crystal characteristics which did not yet have a physicochemical explanation, such as color. The arch-classifier of living things, Carl Linnaeus (1707–1778) had tried to fit minerals into his schemes but with only partial success. Unlike animal and plant species' obvious distinguishing features, the physical and chemical knowledge needed to distinguish mineral "species" simply did not yet exist.

The nineteenth century brought a new approach, both practical and scientific, spurred by advances in the physical sciences. In 1829 the Scottish geologist William Nicol (1771–1851) first peered into the crystalline constituents of minerals by passing polarized light through thinly cut slices, revealing a colorful arrangement of shards, grains, and polygons. Nicol produced polarized light, in which all the light waves vibrate "in the same direction," by passing ordinary light through Iceland spar, an unusually clear type of calcite (a form of calcium carbonate, $CaCO_3$).

This technique proved so valuable that it remains widely used today, as in the image that opens this chapter.

There were also fundamental changes afoot in the understanding of the origin of minerals. James Hutton (page 16) had speculated that many rocks and minerals are forged deep within the Earth by its internal heat and pressure. The Scottish baronet James Hall (1761–1831) performed some "culinary" experiments in which he toyed with mixing, baking, and cooling minerals—cooking sand to make "sandstone" or roasting chalk to make a marble-like substance, and also showing that slow cooling of molten silica (SiO_2) produces crystals, whereas fast cooling yields glass. As was often the case in nineteenth-century geology, it was Charles Lyell (pages 32–33) who drew these strands together to propose that temperature and pressure deep in the crust can convert one mineral into another, potentially quite different mineral. He called this process "metamorphism" and indeed we now consider there to be three main types of rock—igneous, sedimentary, and metamorphic.

In the mid-nineteenth century geologists' attention became focused on how rocks are worn away. In the grand scheme of the planet's history, glaciers are just one cause of erosion, yet much of Europe's terrain just happens to be dominated by their effects—and this is presumably why Europeans became so obsessed with glaciers. It had been known for some

Thomas Moran (1837–1926), *The Grand Canyon of the Yellowstone*, 1893–1901.

Thomas Moran participated in one of the first expeditions to the Yellowstone region of Wyoming.

Preikestolen (Pulpit Rock), Norway, 2016.

Pulpit Rock is a striking and vertiginous result of the carving of Norway's fjords by glacial erosion.

time that glaciers disgorge large boulders, or "erratics," where their "snouts" melt. However, geologists now also realized that those glaciers must once have been much longer, because erratics can be found far down-valley from their "parent" glacier's current snout. In 1837 the Swiss biologist-geologist Louis Agassiz (1807–1873) proposed that the globe had once undergone an *Eiszeit*, or ice-age, when glaciers were longer, much of northern Europe and North America lay under a thick glacial sheet, and that this has left behind a distinctive topography of sharp ridges, U-shaped valleys, and chair-like cwms (also known as corries or cirques). Soon it was realized that glaciers on land sequester enough water to lower sea levels, and by the 1860s geologists had already hypothesized that the weight of glaciers pushes the underlying crust down, and that the crust "rebounds" upward once an ice age has finished.

In contrast, many of the most spectacular landforms of the United States result from erosion by running surface water—and this may explain why American geologists dominated the study of fluvial erosion in the late-nineteenth century. Powell's earlier descent of the Colorado River (page 37), while heroic, had traversed an unusual area—where rivers have cut deep gorges into the Colorado Plateau as the plateau has itself been

"By the 1860s geologists had already hypothesized that the weight of glaciers pushes the underlying crust down, and that the crust 'rebounds' upward once an ice age has finished."

rising. Later, geologists including Clarence Dutton (1841–1912), Grove Karl Gilbert (1843–1918), and William Morris Davis (1850–1934) studied a variety of rivers across the country, many of which are eroding into land which is *not* rising significantly. They discovered, for example, that patterns of river erosion differ between arid regions and wet, vegetated regions. They also hypothesized that rivers erode rock in some places and deposit it at others, depending on the speed of flow and the size of suspended mineral particles. They also characterized how rivers progressively change as they descend to the sea—from cascading torrents in narrow V-shaped valleys, to sluggish, meandering, and "braided" watercourses in wide, flat valleys, estuaries, and deltas.

Around the same time, geologists were beginning to understand other processes which sculpt the planet's surface—the lifting, tilting, distortion, and fracturing of regions of the crust. Geologists had speculated for decades about how mountains form, or "orogeny," but the root cause of mountain uplift was to remain mysterious until the coming of plate tectonics in the 1960s. Some suggested mountains were heaved up by magma swellings

Zabriskie Point, near Death Valley, California, 2015.

This small region of the Amargosa mountain range at the east side of Death Valley is a deposit of lakebed sediments which have been uplifted far above the valley floor, and eroded by surface water into a dramatic landscape.

deep in the Earth, while others proposed a worldwide network of ridges crinkling upward as the planet cooled. Only gradually did the idea of thrust faults gain popularity, in which mountains form when one slab of rock is pushed up and over a neighboring slab. Indeed, although folding of rock beds had been discussed by Hutton in the eighteenth century, faulting was poorly understood until the twentieth.

Looking back, our modern synthesis of the Earth's processes owes a great deal to Hutton. He was a practically minded man who saw soil washed away and understood that it must somehow be replenished; he realized that regions of the land must rise and that something must then bring them low again; he knew that eroded material was not destroyed but merely moved elsewhere. And most of all he speculated about the interconversion of different types of rock by melting, heating, pressure, weathering, and deposition. If geology has one overarching theory, then it is plate tectonics, but if it has one overarching diagram, then it is the "rock cycle." The rock cycle is an attempt to include all the processes to which the planet's minerals may be subject—a very simple version appears below. And we owe much of this global vision to James Hutton.

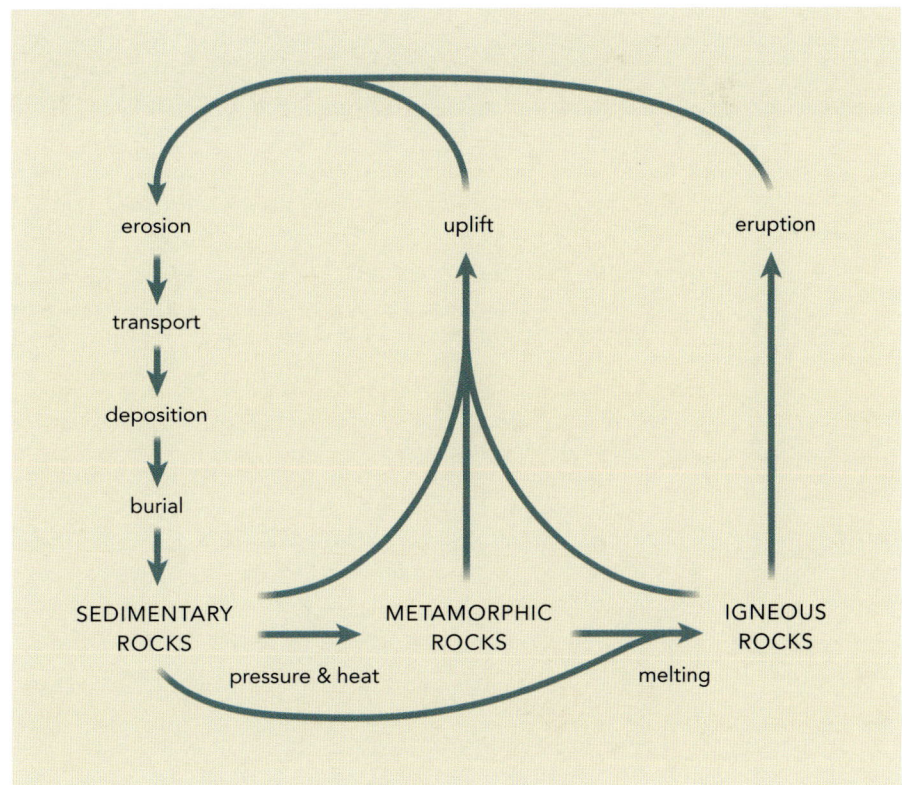

The rock cycle.

This simple version of the rock cycle represents the processes which interconvert the major classes of rock— sedimentary, metamorphic and igneous.

Coast of Bangladesh, Envisat Medium Resolution Imaging Spectrometer, 2003.

The Ganges–Brahmaputra Delta is the largest river delta in the world and has supported a large population for millennia. Spread across the Indian state of West Bengal and much of Bangladesh, its fertile soil now supports around 5 percent of the world's population. Although rivers contain only 0.0002 percent of the world's water, they have a disproportionate effect on its landforms. Deltas form where rivers widen and enter larger bodies of water such as the sea, reducing their flow velocity so that sediment particles sink and are deposited. Throughout the delta there is an unstable balance between deposition, erosion (including by ocean waves), and the tendency of river distributaries to break their banks and cut new channels. River deltas are unusual in that they undergo changes obvious during a human lifetime, and this is why they were the first geological phenomena whose mechanisms and timescales were discussed by the ancients.

The sinuous course of the Büyük Menderes River near Aydın, Turkey.

The Meander River, as it was called in classical times, was important in the ancient world, since it drained a large region of western Asia Minor, and flowed into the Aegean at the Minoan, Hittite, Carian, and Greek city of Miletus. It is now known as the river which gave its name to all sinuous rivers around the world and, coincidentally, Miletus could be argued to be the place where the material philosophy of ancient Greece first arose, eventually forming the foundation of modern science.

Meanders arise because watercourses passing through moderately resistant terrain tend to erode the "outside" of their curves where flow is faster, while depositing sediments on the "inside" of curves where flow is slower. As a result, these curves become more and more exaggerated, leading to a strikingly wiggly course. Eventually, meanders often become so pronounced that they cut through at their "neck" to reroute the stream, leaving an arcuate stagnant "oxbow lake" to one side. For example, satellite photography of the Mississippi River Valley (often erroneously called the "Delta") where the states of Arkansas, Louisiana, and Mississippi meet, clearly shows the meandering river surrounded by old channels and oxbow lakes it has cast aside in the past.

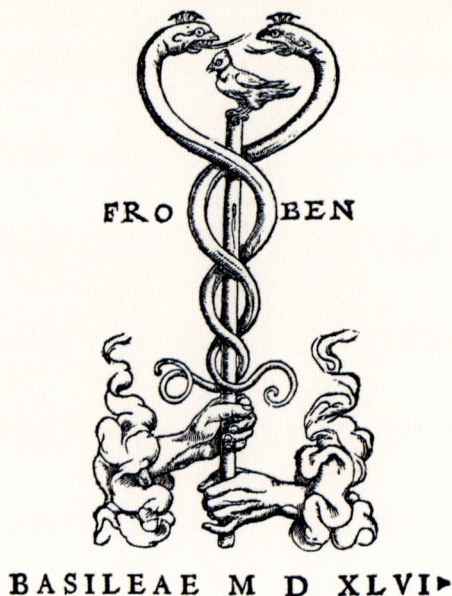

GEORGII AGRI COLAE

De ortu & caufis fubterraneorum   Lib.v
De natura eorum quæ effluunt ex terra   Lib. IIII
De natura fofsilium   Lib.x
De ueteribus & nouis metallis   Lib. II
Bermannus, fiue De re metallica Dialogus.
Interpretatio Germanica uocum rei metallicæ,
addito Indice fœcundifsimo.

FRO BEN

BASILEAE M D XLVI

Georgius Agricola (1494–1555), frontispiece from *De natura eorum quae effluunt ex terra*, 1546.

The German metallurgist-philosopher Georgius Agricola is often called the "father of mineralogy"—and sometimes even of geology itself. Working in his native Saxony and the northern Italian intellectual centers of Bologna and Padua, Agricola combined a practical approach to mining and metallurgy (page 172) with a natural philosophy remarkably unfettered by the authority of classical texts or the Church. In a series of books with names such as *De Natura eorum quae Effluunt ex Terra* (On the Nature of Those That Flow From the Earth) and *De Ortu et Causis Subterraneorum* (On the Origin and Causes of the Subterranean) he was the first to attempt a classification of minerals based on their physical properties. He also discussed how flowing rivers might shape the surface of the Earth, and how water flowing through rocks may deposit mineral veins within them. In addition, he was the first to imply that many gems and minerals are made by the combination of simpler constituents—what we would now think of as a molecular compound being made from the atoms of different elements.

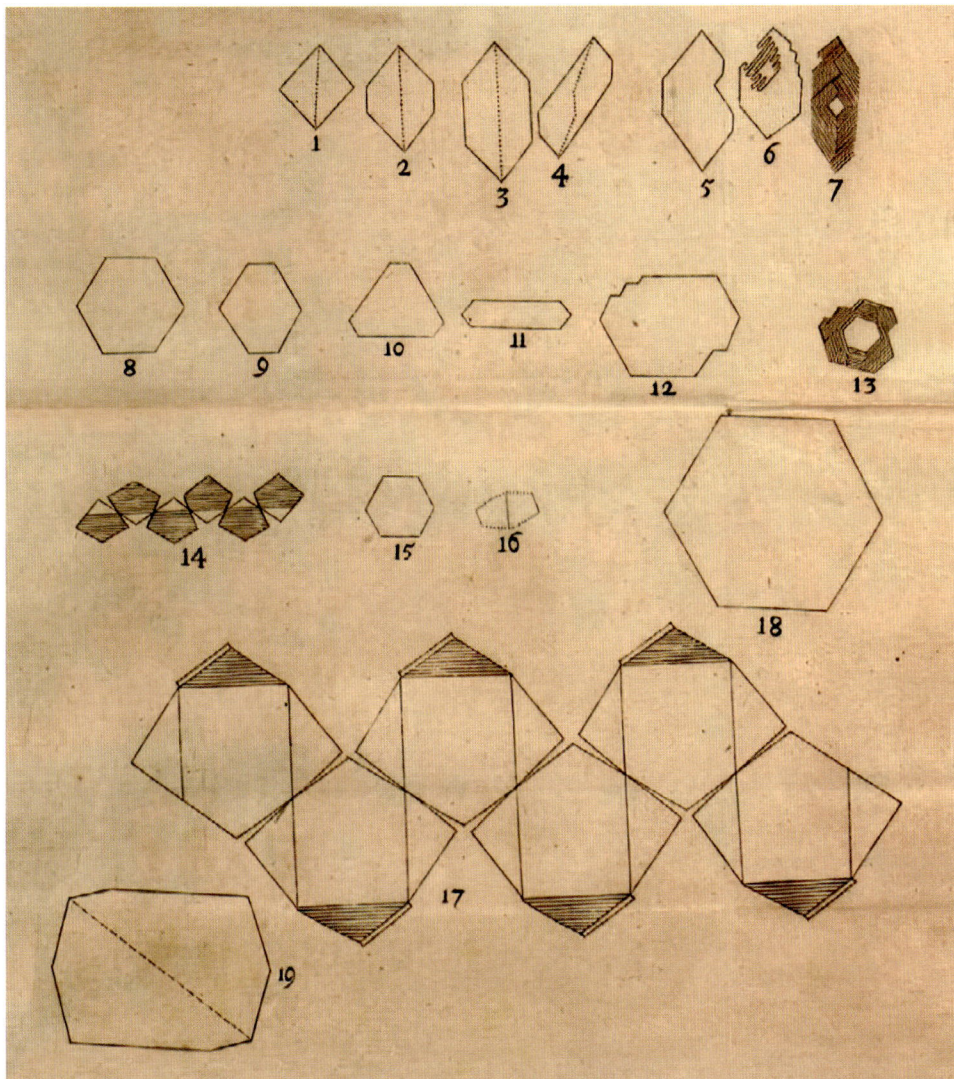

Niels Stensen (1638–1686), also known as Steno, illustration from *The Prodromus to a Dissertation Concerning Solids Naturally Contained Within Solids*, 1669.

As well as his insights into deep time (page 24), the Danish natural philosopher Steno also developed important theories regarding the formation and structure of minerals. In his *De Solido intra Solidum Naturaliter Contento* (*Dissertation Concerning Solids Naturally Contained Within Solids*) he proposed what is now "Steno's Law"—that the angles between the different faces of a crystal are always the same, regardless of the size of the crystal. His phrase "non mutatis angulis" or "without the angles changing" emphasizes that the geometric structure of a crystal is one of its fundamental properties.

Steno also made two other important assertions about crystals—first that they grow not from the inside like living things, but from the outside when fluids deposit additional particles on their facets; and second that a crystal's various faces may grow at different speeds. Thus, Steno established crystals as the fundamental components of most minerals, and that they are "anisotropic"—a crystal's properties of growth, strength, and even light transmission differ, depending on the angle at which it is growing, is compressed, or is illuminated.

In these diagrams, Steno tries to establish the fundamental geometrical elements of crystals, and, in "14" and "17," draws constructions akin to "nets"—which may be folded to create three dimensional polygons.

Carl Linnaeus (1707–1778), *Systema Naturae*, 1768.

The Swedish naturalist Carl Linnaeus developed a system of classifying animals and plants which was to remain dominant until the twentieth century, yet his attempts to fit minerals into this system were somewhat confused. His classification included a very large number of crystal shapes, and he also had strange ideas about the mineral world. Having successfully classified plants based on the shape of their reproductive organs, he then sought the "sexual aspect" of minerals—mixing salty minerals with earthy ones to see what progeny they yielded. And ironically for someone trying to find common themes between the animal, vegetable, and mineral worlds, he did not subscribe to Leonardo and Hooke's idea that fossils are the remains of ancient organisms (see pages 23 and 213)—he thought them no more than aberrant stones.

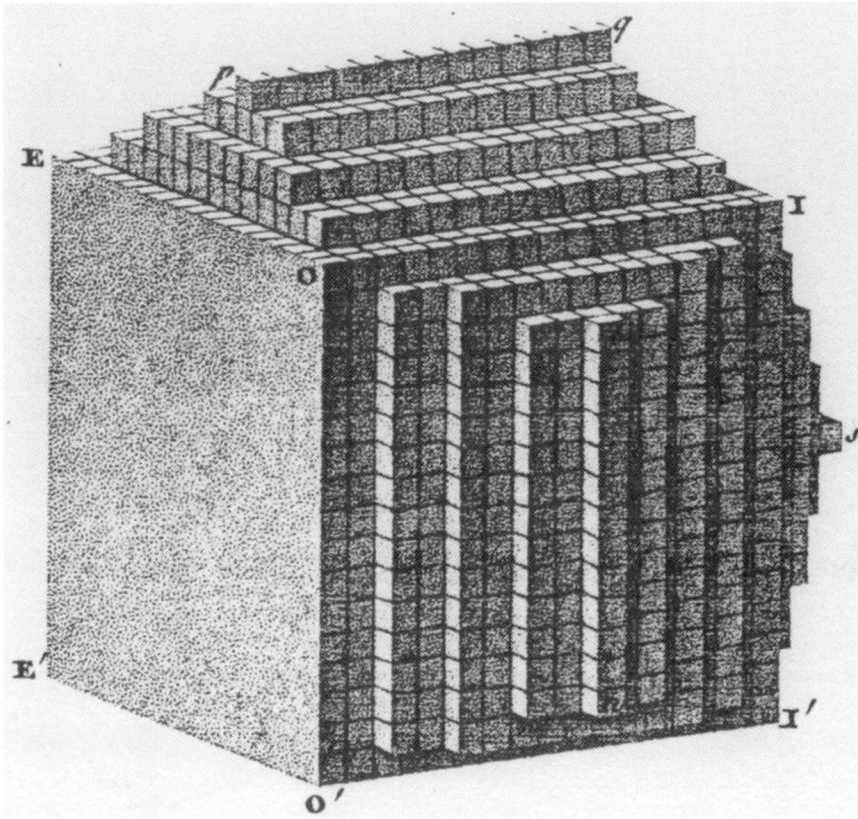

René-Just Haüy (1743–1822), illustration of the *molécules intégrantes* that form pyrite, from *Traité de minéralogie (Treatise on Mineralogy)*, 1801.

The French cleric-mineralogist René-Just Haüy can be said to have started modern crystallography. Fascinated by how large crystals break into smaller crystals which retain the same geometry, he set about measuring the angles between the facets of crystals in all known minerals—or at least those with crystals large enough to measure. He determined that all mineral crystal shapes fall into one of seven classes—including the regular polyhedra (tetrahedron, cube, octahedron, and dodecahedron, with four, six, eight, and twelve sides respectively), "wonky" cuboids called "parallelepipeds," and triangular and hexagonal prisms. Importantly, Haüy also realized that these geometries result from the way crystals' indivisible polygonal units—his *molécules intégrantes*—fit together, as in the cube in the illustration shown here. We now know these polygons actually result from the angles between the bonds linking the atoms in a crystal lattice.

It should be mentioned that the way crystals' chemical formulas are now quoted can be misleading. When a chemist says the formula of carbon dioxide is "CO_2," then it is exactly that—carbon dioxide gas is made up of individual molecules each containing one carbon atom and two oxygen atoms. However, a piece of quartz, despite having the formula SiO_2, does not contain separate molecules each with one silicon and two oxygen atoms. Instead, individual silicon and oxygen atoms are bonded into an extensive geometric lattice in which they are present in the ratio 1:2.

The stirring of scottish stones

JAMES HUTTON (1726–1797)

James Hutton changed our sense of geological time (page 28) but he also changed our understanding of the Earth's processes. Working in the late eighteenth century, he was unusually diligent in backing up his theories with evidence from the varied landforms around him, and he was fortunate to live in the real-life geological textbook that is Scotland. Many of his ideas were published in his 1785 *Theory of the Earth, with Proofs and Illustrations*, and were further disseminated in his friend John Playfair's 1802 *Illustrations of the Huttonian Theory of the Earth*. Most were also later broadcast to the scientific world in the works of Charles Lyell.

Hutton's fieldwork led him to the conclusion that many of the geological phenomena around him were the result of heat and pressure acting on rocks deep within the Earth. For example, at Glen Tilt in the Cairngorm Mountains he found rocks pierced through by veins of granite. Hutton theorized that this could only be the result of that granite once being hot enough to be injected under pressure into the surrounding rock.

He discovered an even more striking example of the effects of heat at the Salisbury Crags, which rise directly to the

LEFT Pink granite veins running through metamorphosed sandstone, 2020.

To Hutton, rocks such as these were evidence that profound changes may be inflicted on rocks long after they first formed—in this case the infiltration of one mineral by veins of another.

RIGHT James Hutton, folded strata from *Theory of the Earth*, 1795.

Dating from the late eighteenth century, this illustration is a demonstration of how great forces within the Earth have caused vast regions of rock to fold and buckle.

east of his native Edinburgh. Here a layer of igneous "whinstone" was forced between layers of sedimentary limestone—a phenomenon called a "sill." Large chunks of limestone seem to have detached from their strata beneath the sill and "floated" upward into the whinstone while it was still liquid—a clear demonstration of how rocks may be altered if they are buried sufficiently deep in the crust.

Hutton was also fascinated by geological folds, where layers of rock are distorted and bent, without much fracturing or crushing. Hutton claimed that only enormous forces within the Earth could throw rock into such spectacular "waves," and that rock becomes "foldable" only if it is exposed to high temperatures and the pressures caused by the weight of overlying crust.

We now know that folding can occur in igneous and metamorphic rocks, but it is easiest to see in the layered strata of sedimentary rocks. It is probably a relatively slow process caused by compression of rock layers from the sides, distortion of layers near faults (page 136), or as soft layers of sediment settle or are compressed unevenly. The *cause* of folding is in fact one of Hutton's rare errors, however—he thought that rock layers crumple when they expand under the influence of heat.

Folds may involve upward flexing of a layer of rock (an "anticline") or downward flexing (a "syncline"), as well as other forms such as steps down or up (a "monocline") or localized basins or domes (page 145). Folds may remain upright, slump to the side, or even partially invert before they are eventually uplifted and exposed at the surface. The study of folds has been of increasing interest since Hutton's time, because once they have formed, they can "trap" valuable ores, petroleum, and natural gas.

TABLE OF FUSIBILITIES.

	Substances.	Original softenid.	Glass softened.	Crystalled softened.
No. 1.	Whin of Bell's Mills Quarry.	40.	15.	32.
2.	Whin of Castle Rock.	45.	22.	35.
3.	Whin of Basaltic Column, Arthur's Seat.	55.	18.	35.
4.	Whin near Duddingstone Loch.	43.	24.	38.
5.	Whin of Salisbury Craig.	55.	24.	38.
6.	Whin from the Water of Leith.	55.	16.	37.
7.	Whin of Staffa.	38.	14¼.	35.
No. 1.	Lava of Catania.	33.	18.	38.
2.	Lava of Sta Venere, Piedimonte.	32.	18.	36.
3.	Lava of La Motta.	36.	18.	36.
4.	Lava of Iceland.	35.	15.	43.
5.	Lava of Torre del Greco.	40.	18.	28.
6.	Lava of Vesuvius, 1785.	18.	18.	35.

ABOVE James Hall (1761–1832), illustration from "Experiments on Whinstone and Lava," *Transactions of the Royal Society of Edinburgh*, vol. 5, issue 1, 1805.

James Hall was a Scottish chemist and UK Member of Parliament. A great friend and supporter of Hutton, Hall respected Hutton's distaste for experimental science, so waited until his death to attempt some mineral cookery in a foundry. He melted igneous rocks from various locations and showed that he could make lava or basalt depending on how fast he cooled them. He also had reasonable success creating rocks akin to marble and sandstone by heating limestone and sand respectively. In this table, Hall records the melting temperatures (in the obsolete "Wedgwood" scale) of rocks from a variety of sources. Hall's geological experiments may be seen as the simple forerunner of today's complex geological phase diagrams (page 146).

RIGHT Abraham Gottlob Werner (1749–1817), "Blues," from *Von den äußerlichen Kennzeichen der Foßilien* (On the External Characteristics of Fossils), 1774.

Although better known as the main originator of the unsuccessful theory of "Neptunism" (page 59), Abraham Gottlob Werner also attempted a comprehensive classification of minerals. Published as *Von den äusserlichen Kennzeichen der Foßilien*, this tome is now mostly remembered for an exhaustive and beautiful attempt to classify the colors of minerals.

We now know that minerals' colors result from several physical processes taking place at the atomic and microscopic levels, as yet unknown in Werner's day. For example, electrons in atoms of transition metals (especially from titanium (element 22) to copper (element 29), see page 39) absorb and discharge packets of energy creating colors visible to the human eye. Also, irregularities in crystal lattices can cause colors, and the fine structure of some crystals is responsible for iridescent or opalescent colors.

BLUES.

Names.	Colours.	ANIMAL.	VEGETABLE.	MINERAL.
Indigo Blue.		Throat of Blue Titmouse.	Stamina of Single Purple Anemone.	Blue Copper Ore.
Prussian Blue.		Beauty Spot on Wing of Mallard Drake.	Stamina of Bluish Purple Anemone.	Blue Copper Ore.
China Blue.		Rhynchites Nitens.	Back Parts of Gentian Flower.	Blue Copper Ore from Chessy.
Azure Blue.		Breast of Emerald-crested Manikin.	Grape Hyacinth. Gentian.	Blue Copper Ore.
Ultra-marine Blue.		Upper Side of the Wings af small blue Heath Butterfly.	Borrage.	Azure Stone, or Lapis Lazuli.
Flax-flower Blue.		Light Parts of the Margin of the Wings of Devil's Butterfly.	Flax-flower.	Blue Copper Ore.
Berlin Blue.		Wing Feathers of Jay.	Hepatica.	Blue Saphire.
Verditter Blue.				Lenticular Ore.
Greenish Blue.			Great Fennel Flower.	Turquois Flour Spar.
Greyish Blue.		Back of blue Titmouse.	Small Fennel Flower.	Iron Earth.

Ball's Pyramid, southwest Pacific Ocean, 2017.

The era of empire-building brought Europeans to far-flung regions and exposed them to natural phenomena they had never seen before. Discovered in 1788, Ball's Pyramid in the southwestern Pacific is one of the most dramatic landforms on Earth—the erosional remnant of the central igneous plug of a 1,000-m (3,000-ft) volcano which formed seven million years ago. It is the tallest sea stack in the world, rising 570 m (1,870 ft) above the water, but is only 300 m (984 ft) wide at its narrowest part.

Ball's Pyramid's unusual structure reflects its complex past. First of all, it is a "batholith"—a mass of igneous rock which accumulated under the surface but did not erupt. Many batholiths develop complex internal structures if they form piecemeal over a long period of time. Also, when molten, their lighter and denser constituents may "differentiate," rising and sinking past each other.

In addition, Ball's Pyramid is unusual in being flanked by extremely steep cliffs which reach far below the current sea level. The landform lies too far south to be protected from erosion by a fringing coral reef—however, the piles of fallen rocks ("talus") which would normally protect such an exposed outcrop seem to have been swept away, perhaps by tsunamis.

Finally, sea erosion is far more intense at the surface than deeper down, yet mysteriously, Ball's Pyramid seems to have been eroded evenly from its visible cliffs down to its base.

Chapeau de gendarme, France, 2021.

Nestled scenically in the Jura Mountains, which gave their name to the Jurassic period, the *Chapeau de gendarme*, or "Policeman's Hat," is an upward "anticline" fold (page 121). Probably deformed into this shape by forces emanating from a nearby fault, the *Chapeau* is, confusingly, made of limestone dating from the Cretaceous, not the Jurassic.

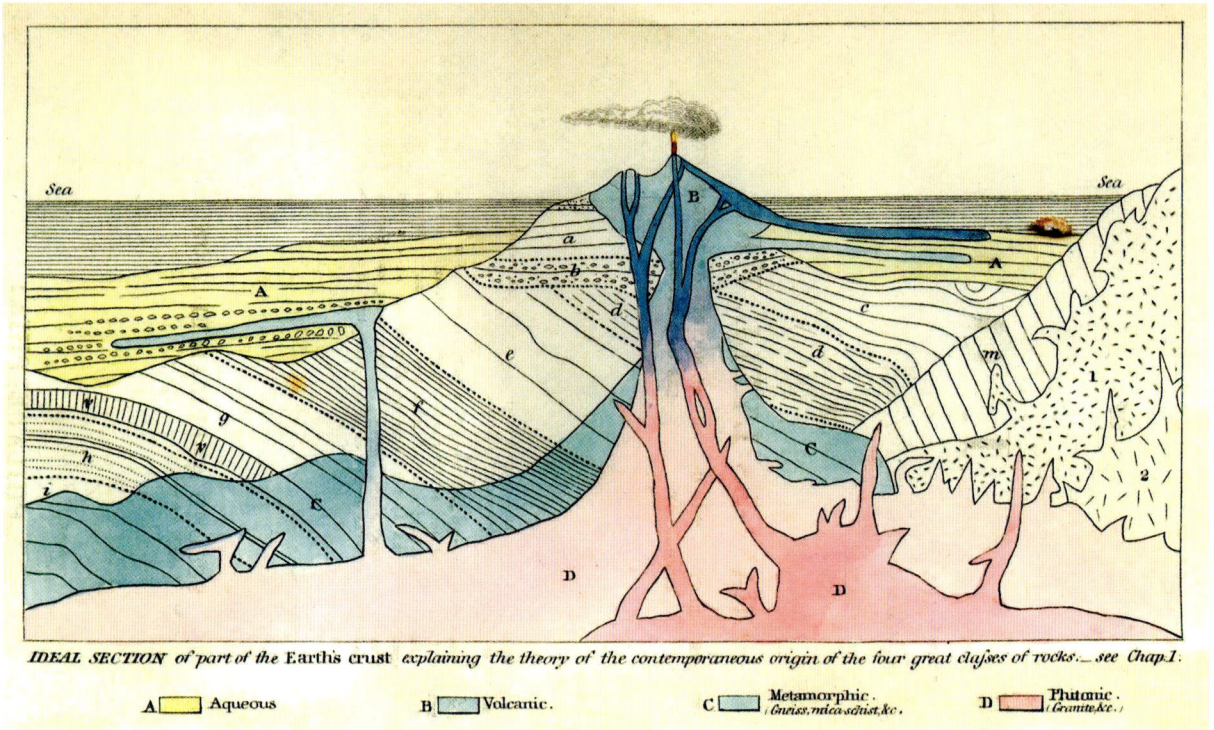

IDEAL SECTION of part of the Earth's crust explaining the theory of the contemporaneous origin of the four great classes of rocks.— see Chap.1.

A ☐ Aqueous   B ☐ Volcanic.   C ☐ Metamorphic. (Gneiss, mica-schist,&c.   D ☐ Plutonic. (Granite,&c.)

Charles Lyell (1797–1875), "Ideal Section of Part of the Earth's Crust," frontispiece from *Principles of Geology*, 1830–33.

The Scottish geologist Charles Lyell made diverse contributions to geology in the areas of volcanology, glaciation, climate, stratigraphy, and deep time (pages 32–33). His books, starting with *Principles of Geology* in 1830, were especially influential in disseminating his holistic, Hutton-like uniformitarian ideas about the Earth's processes. This "ideal section of part of the Earth's crust" is very modern in appearance. It defines the "great classes of rocks" as aqueous, volcanic, plutonic, and metamorphic—although we might now combine the second and third of these as "igneous" (erupted and subterranean, respectively), and rephrase the "aqueous" as "sedimentary."

View across Cwm Idwal from Glyder Fawr, Eryri, Wales, 2012.

In the summer of 1831 Charles Darwin completed his studies at Cambridge University and returned home to Shrewsbury (page 170). Inspired by the geology lectures of Adam Sedgwick he was determined to hone his knowledge over the summer—reading, for example, Charles Lyell's recent *Principles of Geology*. He and Sedgwick left Shrewsbury to conduct a field trip in North Wales, studying the limestone and sandstone strata across the region. They reached the Menai Straits between Anglesey and the mainland and then Darwin struck out on his own, walking south along the coast. This image is of Llyn Idwal (Idwal Lake), now in Eryri Natonal Park, where Darwin noted that despite its elevated location, the lake was surrounded by scattered rocks which contained fossil seashells.

When Darwin joined the HMS *Beagle* expedition, it was as the expedition geologist and zoologist. Although his biological studies of the Galàpagos Islands are more famous, his geological work earlier in the voyage laid the foundations for his famous biological theories. He found beds of marine fossils high in mountains (as he had done in Wales) before he even reached South America, and reported on the lifting of the crust during an earthquake in Chile. He had become a resolute geological uniformitarianist in the mold of Charles Lyell, and it was Lyell who was later to coordinate the co-publication of Alfred Russel Wallace and Darwin's theory of natural selection.

Alexandre-Émile Béguyer de Chancourtois (1820–1886) and Jean-Baptiste Élie de Beaumont (1798–1874), globe showing the relief of the continents and oceans, overlaid by the pentagonal network (*réseau pentagonal*), c. 1850.

A popular figure in his native France, Élie de Beaumont's early theories on the origin of mountains were certainly ambitious. Working at l'*École des mines de Paris*, he suggested that there had been multiple episodes of orogeny in the past, and that they had led to the formation of mountains along a network of circles which gird the world, dividing it into twelve pentagons—*un réseau pentagonal*—as marked on this image with string. Although it was later shown that the world's distribution of mountains does not actually fit this appealing theory, Élie de Beaumont's view of the Earth's surface divided into polygons could be argued to be a slightly over-regularized version of the tectonic plates which now form the basis of modern geology (page 90). Also it is notable that Élie de Beaumont suggested that mountains form along his *réseau* lines as the cooling Earth shrinks—a process soon shown to be inadequate on Earth, but important on Mercury (page 93).

Arnold Escher von der Linth (1807–1872), *View of the valley above the Obersee at Arosa, Switzerland*, c. 1840.

Professor of Geology at the École Polytechnique in Zürich, Arnold Escher von der Linth spent much of his time researching the origins of the Swiss Alps. He observed that in some places large regions of rock seem to have been crumpled or even turned over. He proposed that many mountains are formed by what we now call "thrust faulting," whereby one mass of rock is forcibly heaved over the top of another, presumably as two regions of crust are pushed together. Although some of his explanations for individual Alpine formations have been contested, his general theory is now accepted. The sideways forces which cause thrust faulting often arise where two tectonic plates collide, and the overthrust rocks may slide up a ramp formed by the rocks underneath, or the process may be more complex and lead to dramatic distortion of the colliding regions.

N.º 8. GLACIER du RHÔNE dessiné d'après nature en 1817 par M.ʳ Lardy.

After Monsieur Lardy, in Jean de Charpentier (1786–1855), "Glacier du Rhône dessiné d'après nature en 1817," *Essai sue les glaciers et sur le terrain erratique du bassin du Rhône,* **(Essay on the glaciers and erratics of the Rhône basin), 1841.**

The German-Swiss geologist Jean de Charpentier was the first to suggest that modern glaciers may be shorter than in the past. These slow-flowing rivers of ice pick up large boulders and sweep them down to their "snouts" where they deposit them as they melt. These out-of-place boulders are called "erratics" and are a distinctive feature of glaciers—other erosive forces rarely have enough power to move large boulders. De Charpentier noticed that erratics could be found all along the Rhône valley, deposited far from the glacier's current snout—often near his home near Lake Geneva, in fact. He conveyed his ideas to Louis Agassiz of whose theories they were soon to form an important part (see facing page), but de Charpentier was never to receive the acknowledgment he deserves for changing our view of the Earth's last two and a half million years.

Louis Agassiz (1807–1873), "Hôtel des Neuchâtelois sur la mer de glace du Lauteraar et Finsteraar, côté septentrional," from *Etudes sur les glaciers*, 1840.

Over the course of one night in 1837, Louis Agassiz, Professor of Natural History at Neuchâtel, took up de Charpentier's ideas and formulated the idea of an *Eiszeit* or "ice age" which had once covered much of the Alps with a thick layer of ice. This spilled into long glaciers which carved the U-shaped valleys which now radiate from the mountains, often carrying incongruously diminutive streams along them. Although he could not explain why the climate might fluctuate so wildly, a few years later he expanded his theory to invoke vast continental ice sheets more than 3 km (2 mi.) thick, blanketing much of northern Eurasia and North America.

Glacial periods have come and gone over the last two and a half million years, and there is also evidence of ice ages far older than that. Although ice constitutes only 2.4 percent of the world's water, and most of that is in Antarctica, it has a dramatic effect on the planet's surface and may also drive sudden changes in the evolution and environments of living things. Is it, for example, a coincidence that the most recent retreat of the ice was soon followed by the sudden advent of agriculture, settlements, and civilization?

The above illustration shows intrepid explorers hiking along an Alpine moraine, a ridge of rocky debris deposited at the end of a glacier where it melts. Until the nineteenth century, the vestiges of the erosion, transport, and deposition caused by glaciers were often misinterpreted as evidence of the Biblical flood.

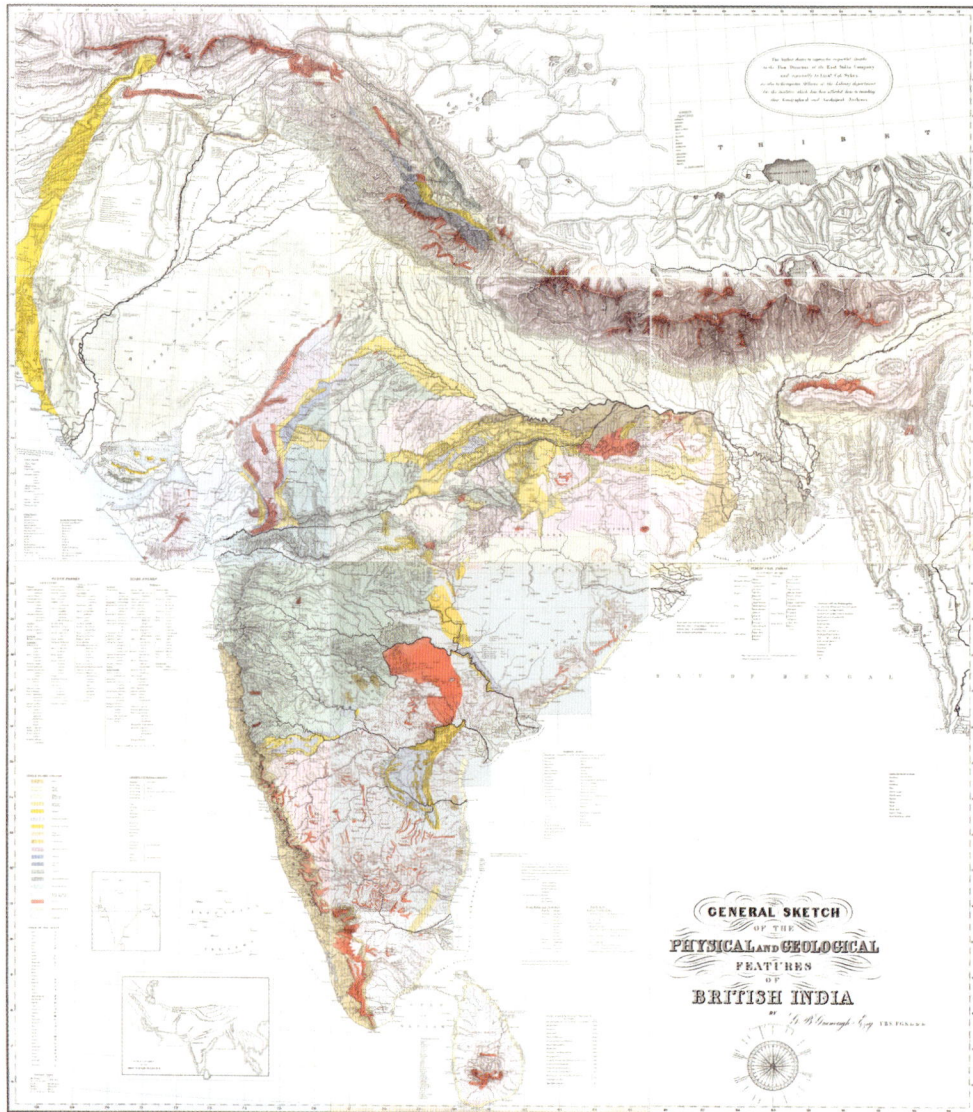

George Greenough (1778–1855), *General Sketch of the Physical and Geological Features of British India,* **1855.**

Goethe's maxim, "What one does not understand, one cannot possess," seems to have been taken to heart by the European colonial powers—and nowhere more so than by the British in India. The British were already well advanced in a titanic seventy-year trigonometric survey of the subcontinent when this geological map was published. George Greenough was the first president of the Geological Society of London and, rather than trekking through mosquito-infested wilderness, he compiled this map from information received from his many correspondents across the width of Britain's most cherished possession.

The Li River running through the karst landscape of Guangxi Province, China, 2019.

In 1893, Serbian geographer Jovan Cvijić (1865–1927) published his *Das Karstphänomen*. "Karst," derived from the Serbo-Croat word *kras*, is a term for landscapes based on limestone rocks, or sometimes other carbonates. Limestone is a form of calcium carbonate ($CaCO_3$) which forms from the compression of mollusk shells and coral. The characteristics of karst landscape result from limestone's solubility in rainwater acidified by carbon dioxide. Limestone can be dissolved away to leave a variety of dramatic forms including gorges, rock arches, caves, springs, and sinkholes—and much of the water in these regions runs underground in a network of eroded channels. Sinkholes may become the location of isolated "islands" of forest and animal life, protected from human intrusion by their precipitous sides. More than 10 percent of East Asia is karst, as is, in Europe, the land above Trieste.

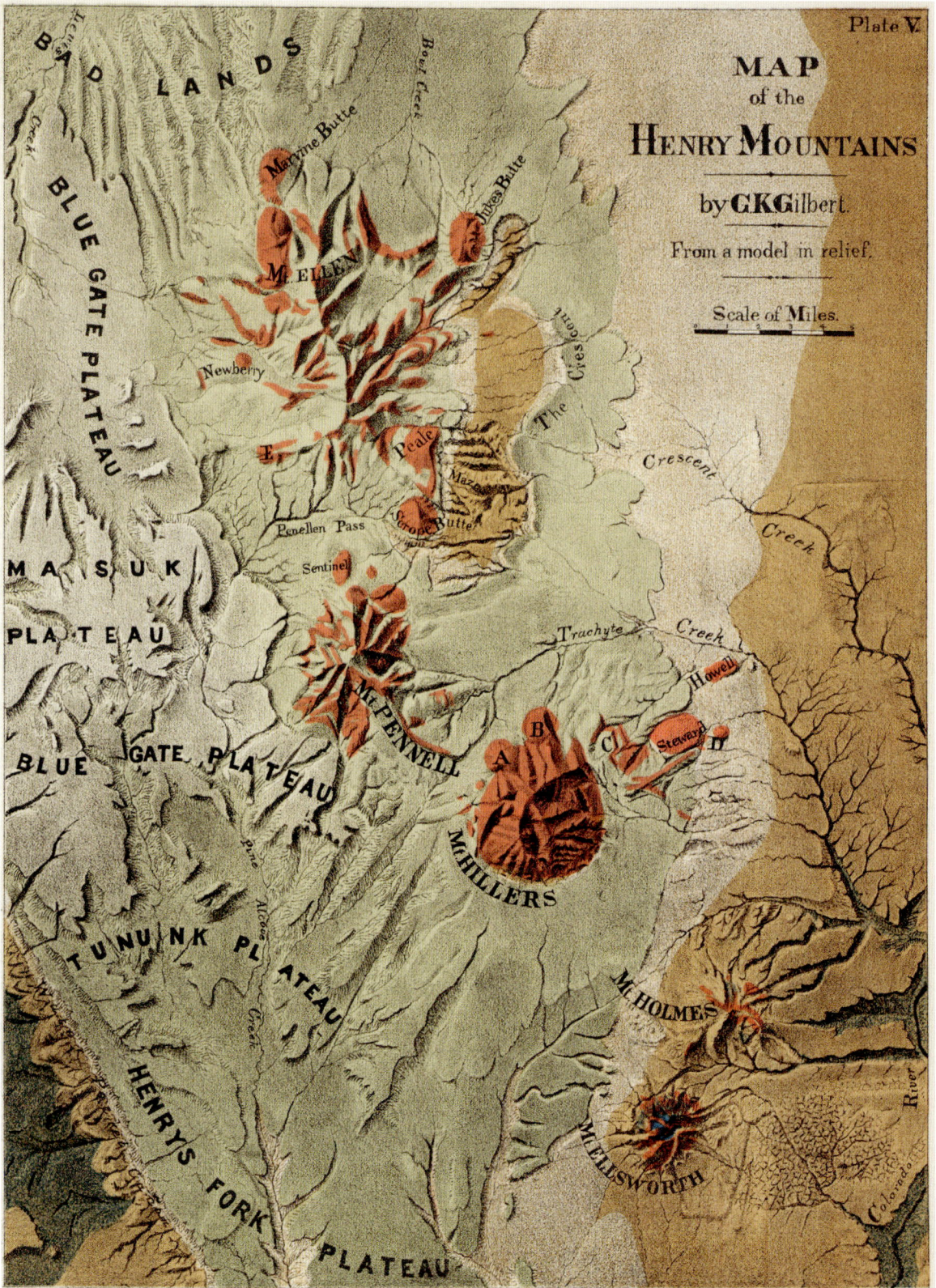

Plate V.

MAP
of the

HENRY MOUNTAINS

by C.K.Gilbert.

From a model in relief.

Scale of Miles.

BAD LANDS

BLUE GATE PLATEAU

Marvine Butte

Jukes Butte

Mt. ELLEN

Newberry

E.

Peale

Mazon

The Crescent

Crescent Creek

Penellen Pass

Scrope Butte

Sentinel

MASUK PLATEAU

Trachyte Creek

Howell

Mt. PENNELL

A   B   C   Steward   D

BLUE GATE PLATEAU

Mt. HILLERS

Pine Alcove Creek

TUNUNK PLATEAU

Mt. HOLMES

River

Sheep Creek

HENRYS FORK PLATEAU

Mt. ELLSWORTH

Colorado

LEFT Grove Karl Gilbert (1843–1918), "Map of the Henry Mountains," from *A Report on the Geology of the Henry Mountains*, 1877.

In the 1870s John Wesley Powell (page 37) commissioned Grove Karl Gilbert to survey a remote area in southern Utah. This region was one of the last parts of the United States to be mapped, and Gilbert named the peaks which had sparked Powell's interest as the Henry Mountains. Gilbert proposed that these mountains had formed in an unusual way, as rising domes of subterranean magma raised blister-like swellings on the surface. He claimed that these had then been partially eroded by running water to yield today's dramatic skyline. Gilbert became fascinated by the action of water on landscapes over geological time, and how watercourses pick up particles of rock in one place and then deposit them in another. He also noticed that rivers and streams affect the surrounding terrain differently depending on whether or not its upper layers are stabilized by vegetation.

ABOVE The Susquehanna River cutting through the folds of the Valley-and-Ridge province of the Appalachian Mountains, 2019.

William Morris Davis (1850–1934) was enormously influential in the development of American geology. He studied geology and geography at Harvard, and initially trained as an engineer. Working especially in Pennsylvania, Davis developed a model of how rivers develop over time—somewhat inspired by the new theories of animal evolution. According to this model, regions of crust are uplifted and then are gradually eroded away by flowing water. Over time, the rivers and streams evolve from fast-flowing watercourses in steep-sided V-shaped valleys into broader rivers in flatter valleys, with meanders and "braided" sections, deltas, estuaries, and mudflats. Thus, the tendency is for high ground to "evolve" toward a "base level" of a flat, eroded plain traversed by sluggish rivers—in preparation, perhaps, for the next cycle of uplift to occur.

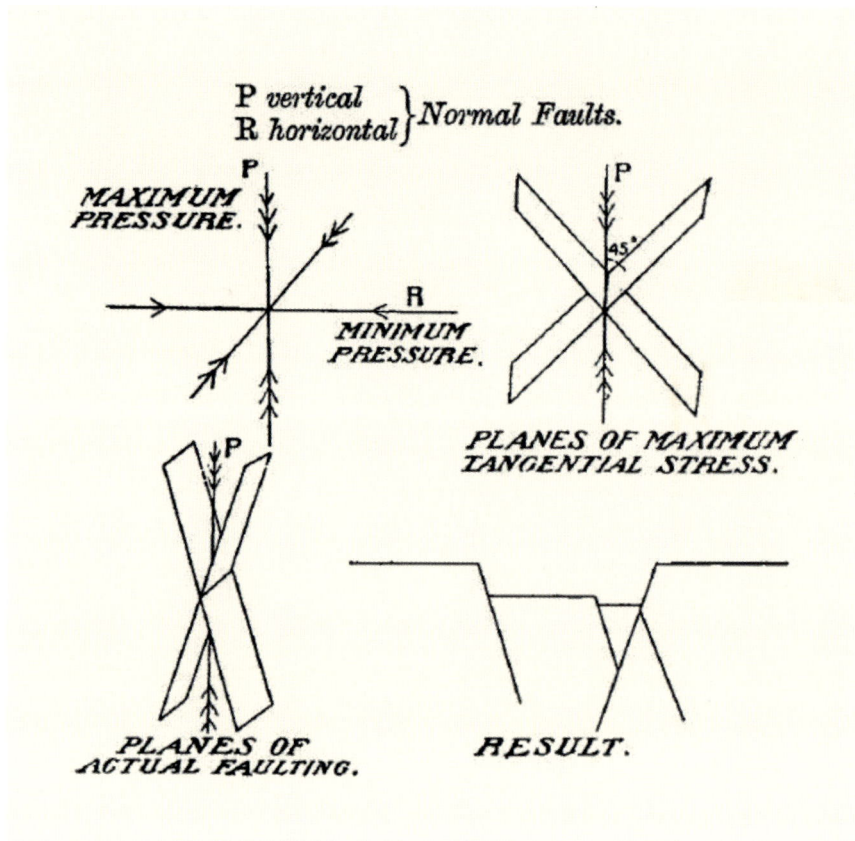

Ernest Anderson (1877–1960), illustration from "The Dynamics of Faulting," 1905.

Although geological folding had been discussed since Hutton's time, faulting was
not well characterized until the early twentieth century—partly by the Scottish
geologist Ernest Anderson who systematically classified the different types of
faulting and the forces which cause them. Faulting occurs when two regions of
rock fracture from each other, and then move relative to each other along that
fracture plane. For example, a "strike-slip" fault occurs when the two regions slide
horizontally "sideways" alongside each other—such as at the San Andreas Fault
(page 77).

A small graben along the Zanjan-Tabriz Highway, Iran, 2020.

Regions of rock may also move vertically relative to each other, often sliding along an oblique fracture plane. Sometimes one portion slides "down" the slope—a configuration called a "normal fault." When, as in the picture above, a region slips down at both ends along two normal faults, the resulting depressed chunk is called a "graben"—from the German for "ditch." A region left elevated between grabens is called a "horst"—meaning "heap."

In contrast, if one region rides "up" the fracture slope, the result is a "reverse fault." If this occurs where the diagonal fault is at a shallow angle, it is called a "thrust fault" like those which Arnold Escher von der Linth (page 129) suggested caused the piling-up of the Alps.

Flattened trees, near the site of the Tunguska Event, Russia, c. 1927.

As early at 1892 Grove Karl Gilbert (page 135) correctly hypothesized that lunar craters are the result of impacting objects—and that these impacts explain craters' shapes, as well as the rays of pale ejected material which radiate from them. He did not, however, think that such events occurred on Earth, claiming that what is now called "Meteor Crater" in Arizona was created by a volcanic explosion.

However, in the morning of June 30, 1908, an event occurred in Siberia which changed attitudes to the possibility of large extraterrestrial impacts on Earth. In the region of the Podkamennaya Tunguska River, a huge fireball was seen, brighter than the sun, the ground shook, and rumbling sounds were heard up to 1,400 km (870 mi.) away. Over the coming nights, strangely colored clouds and glowing skies were seen across Eurasia. Soon, Russian scientists far from the site suggested that the event was caused by the impact of an extraterrestrial object.

The site was not visited by outsiders until 1927, when 2,000 sq km (800 sq mi.) of forest was found to be flattened. However, there was no crater, and no incriminating chunks of extraterrestrial impactor. Over the decades this lack of evidence led to a wide range of theories—memories of Krakatoa inspired theories of volcanic explosions, Cold War paranoia elicited stories of alien nuclear weapons, and from the more exotic realms of physics came the theory that a small black hole had hit the planet. Currently, the prevailing theory is that it was indeed an extraterrestrial body which entered the Earth's atmosphere that day in 1908, but it exploded approximately 8 km (5 mi.) above the surface.

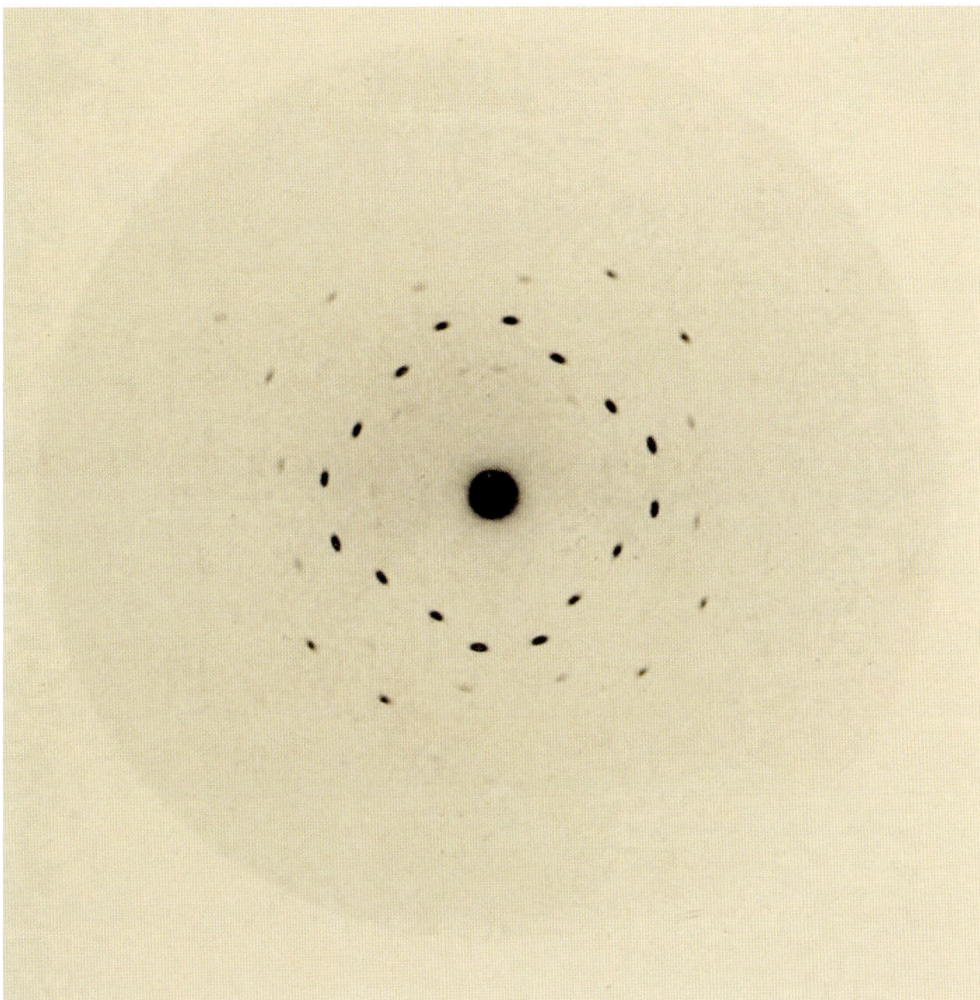

Max von Laue (1897–1960), illustration showing diffraction of X-rays by zinc sulfide, from "Interferenz-Ersheinungen bei Röntgenstrahlen," 1912.

It is remarkable that as early as 1912 a technique was developed which allowed scientists to peer into the actual atomic structure of crystals—X-ray crystallography. It had been known for some time that visible light can be "diffracted"—effectively bent around corners—and that if many tiny, regularly-spaced "obstacles" are put in the way, light is diverted and split into a rainbow of colors. For example, diffraction is the mechanism by which compact discs reflect a spectrum of colors.

Early in the twentieth century, physicists wondered if diffraction could be used to determine the atomic structure of crystals, yet the wavelength of visible light is far too long for it be diffracted through the minuscule gaps between the atoms in a crystal. However, the discovery of X-rays in 1895 by Wilhelm Röntgen (1845–1923) was to change all that. Like visible light, X-rays are a form of electromagnetic radiation, but they have a much shorter wavelength—a wavelength, in fact, which the German physicist Max von Laue realized *would* be diffracted by the atoms in a crystal. He passed X-rays (*Röntgenstrahlen*) through crystals of copper sulfate ($CuSO_4$) and zinc sulfide (ZnS) and produced spotty images on radiographic film. Each crystal type deflected particular amounts of X-rays in particular directions, in a configuration which allowed its atomic structure to be deduced.

Even today, X-ray crystallography remains the definitive way to determine the atomic structures of mineral crystals and biological molecules—in the 1950s it was the method used to characterize the double-helix structure of the DNA molecule which carries our genetic information.

Extraordinary vigorous subfluvial action

J HARLEN BRETZ (1882–1981)

In the dry eastern regions of Washington state, the Columbia Plateau presents one of the strangest landscapes in the world. The "Mighty" Columbia River, the largest river draining the western slopes of the Americas, carves its way through a wild landscape of gorges, scoured bedrock, precipices, gravel mountains, plunge pools, and great tracts of rippling land strewn with isolated boulders.

Away from the Columbia, there are wide dry valleys or *coulees* through which water once drained (from the Canadian French *coulier*, "to flow"). Moses Coulee remains in its original dry state, but the Grand Coulee is now flooded by water for crop irrigation.

It had been assumed that only glaciers can cut sweeping valleys and deposit erratics, and that the erosive effect of flowing water is more sedate—slowly carving V-shaped or flat valleys. As a result, the channeled scablands of Washington were thought to be nothing more than old routes of the winding Columbia. However, in the early 1920s the geologist J Harlen Bretz proposed that something altogether more violent had happened on the Plateau.

The Columbia Plateau is a huge slab of igneous basalt around 2 km (1¼ mi.) thick,

LEFT The Grand Coulee, Washington State, 2015.

RIGHT Dry Falls, Washington State, 2015.

spewed out between seventeen and six million years ago. Bretz proposed that its modern-day surface topography was created by a series of cataclysmic flood events which tore across it from east to west. He argued that only sudden tumultuous events could explain the area's unusual topography.

According to his theory the coulees were temporary drainage channels carved over a matter of days as the inundation from the east overtopped the banks of the nearby Columbia and swept across the plateau to the sea. Indeed, a great subsea canyon has recently been discovered, gouged into the continental shelf by surges of water disgorged into the Pacific at the mouth of the Columbia.

The most dramatic evidence of these floods is Dry Falls, located partway down the Grand Coulee where, for a few days, ten times the combined flow of all the world's rivers tumbled over a waterfall 120 m (390 ft) high and 5.5 km (3½ mi.) wide. As their name suggests, the falls are now dry, except for the lakes that have formed in their great plunge pools. It is hard to believe, but the flow over the falls may have been so great during the floods that the surface water level did not visibly change here, despite the huge drop.

Bretz's theory was understandably controversial, partly because he could not explain where the floods came from, and why they were so sudden and formidable. This problem was solved by Joseph Pardee (1871–1960), who realized that a huge glacial lake, Lake Missoula, had formed over much of what is now western Montana around fifteen thousand years ago, held back by ice dams 600 m (1,970 ft) high. These ice dams repeatedly failed, unleashing the great floods. Each flood lasted only a matter of days, but several occurred over a period of two thousand years.

Rock fissured by the freeze-thaw process, Petermann Island, Antarctica, 2010.

Water has unusual physical properties which explain many features of the world around us. Most liquids shrink and become denser as they get colder, until they freeze—when they become denser still. However, although warm water does indeed get denser as it cools, below 4°C (39°F) it *expands*, becoming less dense. And as water freezes, its crystal structure becomes so "spread out" that its volume increases by approximately 10 percent. The expansion of freezing water can be a powerful erosive force—water can seep into cracks in rocks, and then prize them apart as it freezes, as in the picture above.

In 1927, the Norwegian adventurer-scientist Finn Malmgren (1895–1928) wrote *On the Properties of Sea-ice*. Because ice is less dense than water, sea ice floats on the oceans rather than sinking, and 10 percent of it protrudes above sea level. This also explains why climate-change-induced sea-ice melting will not raise sea levels in the future—the floating ice will "shrink" as it melts, and perfectly "fill in" the volume of water it previously displaced. Things may be different, however, elsewhere in the solar system (pages 246–247).

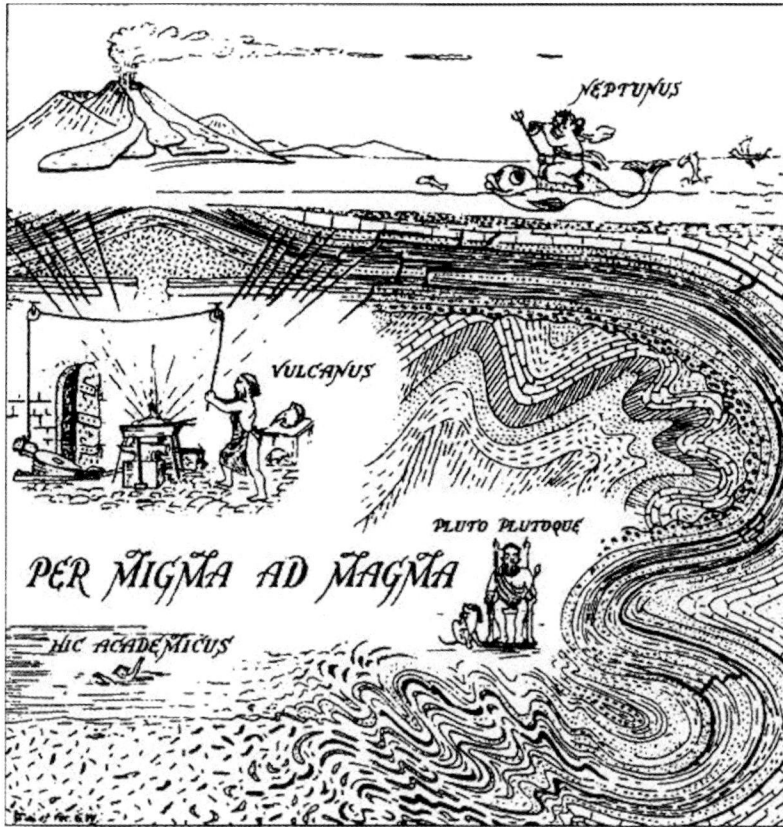

Dr. Gilbert Walton, "The Rocks Display'd," frontispiece from H. H. Read (1889–1970), *The Granite Controversy*, 1957.

For such a common rock, granite has been strangely controversial. Back in the eighteenth and nineteenth centuries granite was at the center of the debate between Neptunists and Plutonists—with Werner's supporters claiming that the Earth's granite crystallized from a primordial ocean, and Hutton's arguing it was produced by the solidification of molten rock.

Granite is a coarse-grained (hence the name) rock rich in silica (SiO_2) with an admixture of the oxides of aluminum and other metals (Al_2O_3, Na_2O, K_2O, and so on) which makes up a large amount of continental crust. It is thought to have formed underground as silica-rich igneous material cooled slowly, sometimes in the presence of water, and it may also be formed by modification of igneous basalt derived from oceanic crust. Not much granite seems to have been produced in the last 600 million years and this, along with the fact that the processes which produce granite are not fully understood, have led some to claim that we do not really know why there is so much of it.

The Richat Structure, Mauretania, 2001.

As its noncommittal name implies, the Richat Structure was named at a time when it was one of the most mysterious large-scale formations on the planet. Discovered in the 1930s, it has been variously claimed to be the site of an extraterrestrial impact, the location of Atlantis, or a megastructure built by aliens. It is in fact a dome—a region of rock elevated above its surroundings when a blob of magma rose beneath it. Domes are a localized form of gentle folding, and as the bulging layers of rock are eroded away, concentric circles of different strata are exposed. Most rock has been eroded from the center of the Richat Structure, so the oldest exposed rocks are found there, with concentric rings of progressively younger rocks arrayed around it.

THOLEIITE + 5% H$_2$O

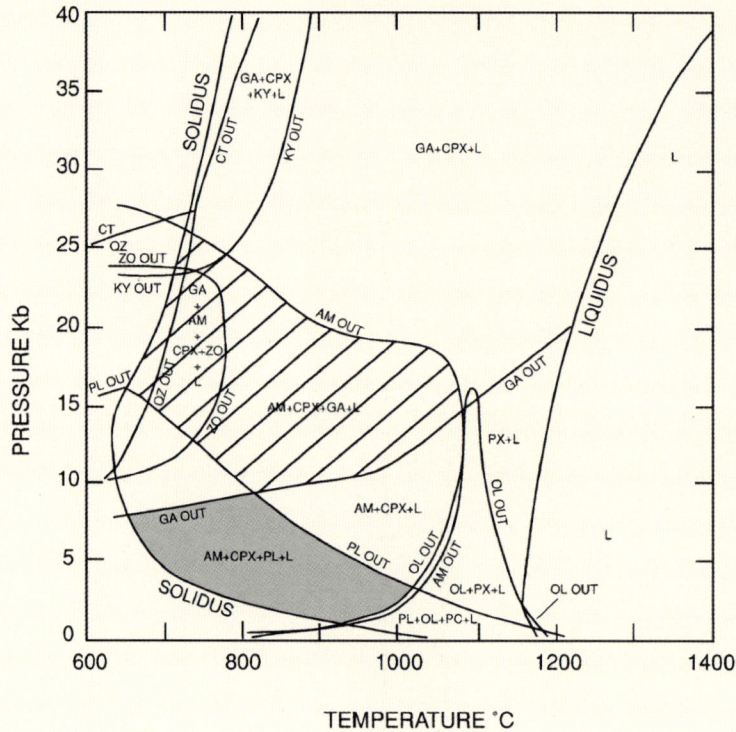

PRESSURE Kb

TEMPERATURE °C

SOLIDUS · CT OUT · KY OUT · GA+CPX +KY+L · GA+CPX+L · L · CT · QZ · ZO OUT · KY OUT · GA + AM + CPX+ZO + L · PL OUT · QZ OUT · ZO OUT · AM OUT · AM+CPX+GA+L · LIQUIDUS · GA OUT · PX+L · GA OUT · AM+CPX+L · OL OUT · OL OUT · L · AM+CPX+PL+L · PL OUT · OL OUT · AM OUT · OL+PX+L · OL OUT · SOLIDUS · PL+OL+PC+L

Lagabrielle et al., pressure-temperature diagram showing experimental phase relations for dry and wet basalts, from "Magmatic-Tectonic Effects of High Thermal Regime at the Site of Active Ridge Subduction: The Chile Triple-Junction Model," *Tectonophysics*, vol. 326, pp. 255–68, 2000.

We now realize that a coherent classification of minerals was simply impossible for early geologists. Minerals have turned out to be controlled by complex physicochemical rules which determine what minerals consist of, and how they change under different conditions.

Today, these phenomena are often visually represented using *phase diagrams*— graphical representations of different forms of minerals under differing conditions and at different times. The square diagram here is one of the simplest types, illustrating the varying forms that basalt can take at different temperatures (horizontal axis, hotter on the right) and pressures (vertical axis, higher pressures at the top).

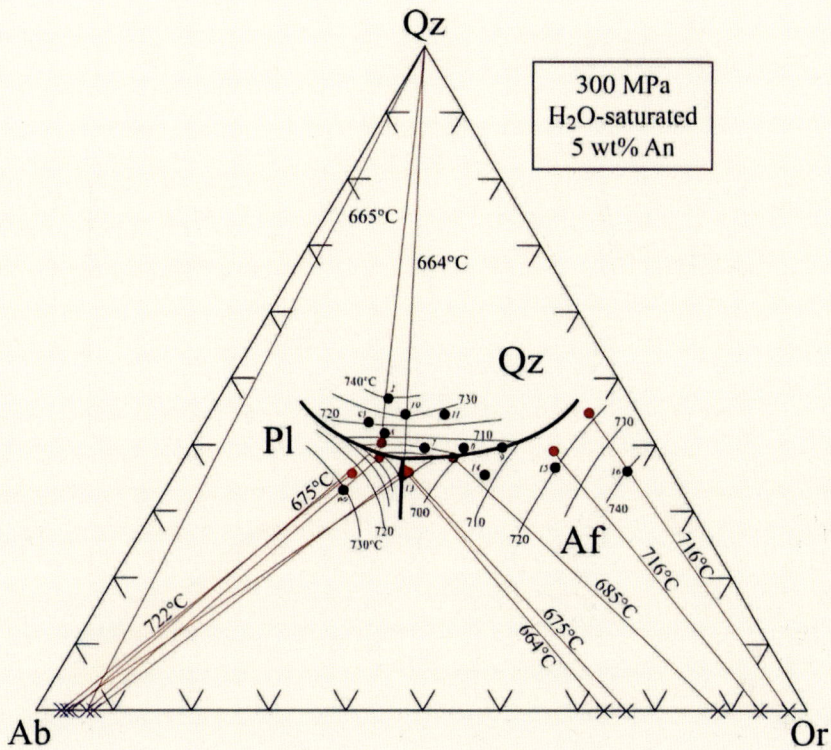

Pichavant et al., liquidus phase diagram for the Qz-Ab-Or system, from "Effect of Anorthite on Granite Phase Relations: Experimental Data and Models," *Comptes rendus—Geoscience*, vol. 351, pp. 540–50, 2019.

Minerals sometimes behave in unexpected ways. Heat and pressure may convert them into new forms, and sometimes these new forms remain stable when exposed at the surface of the Earth, whereas others may be unstable and gradually transform into yet another mineral. When they melt, some minerals remain pretty much the molten form of the same mineral, whereas other minerals fundamentally change their make-up when they melt. Some molten minerals coexist stably with other minerals in solid form, while others do not.

And most complex of all, many rocks are made up of a mixture of different minerals, all interacting with each other—minerals often behave differently simply because they are mixed with other minerals. For example, the triangular diagram illustrates complex interactions in a granite-like mixture of quartz ("Qz," SiO_2), albite ("Ab," $NaAlSi_3O_8$), and orthoclase feldspar ("Or," $KAlSi_3O_8$).

Shorelines and riverbanks of the solar system

IS THERE LIFE ELSEWHERE?

Geology's Copernican moment came quite late—with confirmation that Earth's geological and hydrological processes are similar to those elsewhere in the Solar System. In 1971, Mariner 9 reached Mars and became the first manmade object to enter orbit around another planet. The probe sent back unmistakable photographic evidence of surface erosion by running water.

Although Mars' surface has been very dry for the last 3000 million years, it retains evidence of ancient valleys eroded by sinuous and meandering watercourses, with midstream islands and oxbow lakes (page 115). And where these ancient streams once drained into depressions in the Martian crust, layers of sediment are found, presumably deposited at the bottom of lakes long since vanished.

More recent exploration has shown that Mars is not completely arid. Hundred-meter-long streaks of dark mineral salts descending Martian slopes—"recurring slope lineae"—are evidence of briny water exuding from the surface and flowing downhill. Although Mars' surface is too cold for pure water to be liquid, dissolved salts lower its melting point so that it can sometimes exist as a liquid on the Martian surface—and indeed the lineae are more prominent during warmer seasons.

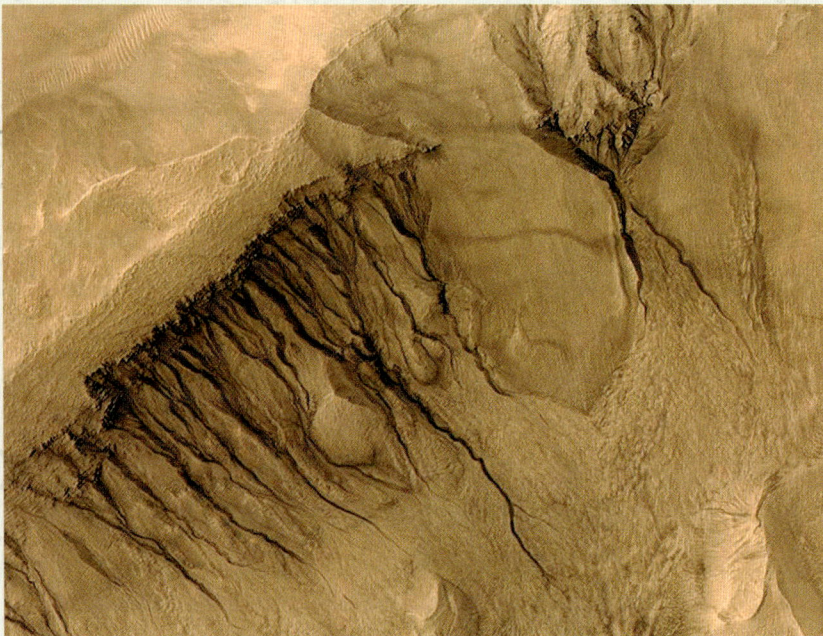

Gullies probably formed by water and debris flows, 2003.

This is an image of part of the wall of a small Martian impact crater, taken by the Mars Global Surveyor spacecraft. Similar channels on Earth are usually created by the action of flowing water.

The lineae are probably the surface manifestation of much larger subterranean flows, and in 2024 evidence emerged of even larger bodies of water under the surface. Seismological data have indicated that large bodies of *liquid* water exist 10–20 km (6–12 mi.) below the surface. And where there is water, there is the possibility of life—after all, many of Earth's microbes live below the surface, sometimes even within its rocks (page 243).

It is now also clear that a celestial body does not even need liquid water to host Earth-like liquid processes. In 1980 the Voyager spacecraft flew past the Saturnian satellite Titan and showed that the moon has some unusual chemistry taking place in its atmosphere—based on hydrocarbons similar to those found in natural gas deposits on Earth. And importantly, Titan's surface temperature is in the range where those hydrocarbons can exist in liquid form.

In 2004 the NASA Cassini probe entered orbit around Saturn and stayed there for thirteen years. It used a variety of detectors to peer through Titan's haze and discovered large lakes of methane and ethane (CH_4 and C_2H_6). There is also evidence of remarkably Earth-like liquid-cut channels leading into these lakes, even though the moon's 180°C (356°F) surface temperature means that any actual water would be brittle solid.

Indeed, it is now thought that Titan has liquid cycles analogous to the water cycle on Earth, but slower and based on the liquid hydrocarbon molecules. Benzene (C_6H_6) may fall from the sky as snow and settle as sludge on the ground or in lakebeds. Hydrocarbons may solidify on the great lakes of Titan, but unlike water ice (page 142) any tiny "bergs" would sink to the lake floor.

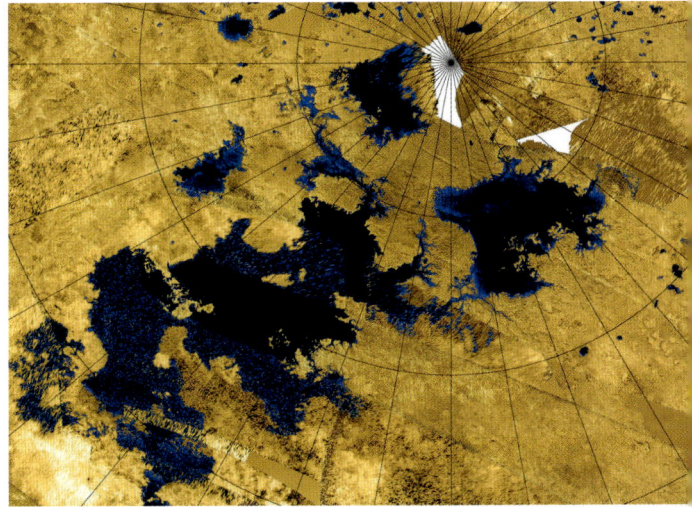

TOP **The lakes of Titan, composite image 2004–13.**

This is how the north pole of the Saturnian moon Titan would appear, were its thick atmosphere stripped away— the dark blue regions are the satellite's hydrocarbon lakes.

ABOVE **Jezero Crater, Mars, 2011.**

Imaged here by NASA's Mars Reconnaissance Orbiter, Jezero Crater was chosen as the 2021 landing site for the later Perseverance Rover. At the edge of the 45-km (28 mile) wide crater lies a formation thought to represent an ancient river delta (compare with page 114). The colors in this image reflect the mineral composition of the surface rocks, as measured by the Orbiter's onboard instruments.

ABOVE **Goosenecks, Utah, 2017.**

RIGHT **Umpire Rock, with glacial striations, Central Park, New York, 2009.**

The influence of erosion on the Great American Landscape spans the nation from its most remote gorges to its greatest metropolises. A state park since 1962, Goosenecks is a striking example of an entrenched meander—a series of dramatic sunken bends in a river's course which result from an unusual, almost paradoxical, set of conditions. These meanders first formed when the San Juan River, a tributary of the Colorado, eroded a shallow, lazy, sinuous course across the relatively flat Colorado Plateau. Rivers on flat plains do not usually cut deep valleys, but the region across which the San Juan flowed was then slowly uplifted, at the same rate as

the river was eroding downwards. As a result, the river gradually cut deeper and deeper as the land rose around it, "freezing" in precipitous form its shallow ancient meanders.

Although it may not seem like it at first site, much of New York sits on a glacial landscape—the bedrock of Central Park, for example, bears the scars left as rocks were dragged over it during a period of glaciation twenty-one thousand years ago.

Nearby, Long Island has formed around a series of terminal moraines—ridges of rubble dumped as the glaciers creeping across New England melted at their southern snouts. In the west of the island the ridge runs along the north coast, facing Long Island Sound, whereas in the east the ridge runs along the center. The highest parts of Long Island are more than 100 m (330 ft) high, demonstrating the sheer scale of glacial deposition.

LEFT **Cave of Crystals, Naica, Mexico, 2006.**

The Naica lead, silver, and zinc mine in the Mexican state of Chihuahua was opened in 1828—in a region where a limestone landscape had cracked and fissured when a magma dome rose beneath it twenty-five million years ago. In 1910 the large Cueva de las Espadas (Cave of Swords) was discovered, from whose walls hung meter-wide crystals of transparent gypsum (hydrated calcium sulfate, $CaSO_4.2H_2O$). In 2000, miners stumbled upon the even more spectacular Cueva de los Cristales, containing gypsum crystals more than 10 m (33 ft) long and 3 m (10 ft) wide.

Large crystals form only under certain conditions—and those conditions must persist for extremely long periods of time. It is thought that heat from the underlying magma warmed subterranean water so that it partially dissolved the surrounding limestone. Saturated with calcium and sulfate, it started to slowly deposit $CaSO_4$ crystals in the water-filled cave. The gigantic crystals were only discovered because nearby mining had led to the draining of the cave, and indeed they are once again submerged and inaccessible now that the mine has closed.

ABOVE **Klerksdorp sphere, South Africa.**

Discovered in South Africa's northwest province, the Klerksdorp spheres are disconcertingly artificial-looking spheroids found embedded in 3000-million-year-old rocks. Often claimed to be relics of an ancient alien visitation or a long-lost terrestrial civilization (page 231) these spheroids with their circumferential inscribed grooves have turned out to have a more prosaic explanation. Geological studies published in 2007 and 2008 explained that they are conglomerate concretions—dollops of hematite (Fe_2O_3) or pyrite (FeS_2) which were deposited in the midst of unusually fine and homogenous particles, so that they could grow evenly in all directions, yielding near-spherical masses. Even the "surface" grooves have a natural explanation—they are no more than the external evidence of the spheres' internal layered structure.

ABOVE Impressions of raindrops preserved on a bedding plane, the Taller Mecanico formation, Spain, 2008.

It is hard to believe that something as evanescent as a raindrop might fossilize, but many geologists believe they do—and some raindrop fossils are billions of years old. Most fossilized raindrops probably form when rain falls on freshly deposited fine volcanic ash, their imprints are quickly covered over by another layer of fine ash, and then compacted into rock over a long period of time. Much later, that rock is exposed at the surface and partially eroded to reveal the ancient drop marks once more. Raindrop marks may seem unlikely to fossilize as efficiently as something substantial like a dinosaur, but there have been many, many times more raindrops than dinosaurs over Earth's history.

As far back as the mid-nineteenth century Charles Lyell (pages 32–33 and 126) suggested that the size of fossilized raindrops could be used to estimate the density of the planet's ancient atmosphere, because the maximum size of a raindrop depends on the density and constituents of the atmosphere through which it falls. Indeed, a 2012 study showed that atmospheric pressure was between one half and double its current levels 2,700 million years ago.

RIGHT Raditladi impact basin, Mercury, false-color image, 2011–2015.

For a planet once assumed to be an inert sunbaked wasteland, tiny Mercury has proved surprisingly important in setting Earth's geological processes into a wider context. This torrid world was once thought to be largely free of "volatiles"—substances which when heated melt and evaporate away into space—but it now appears that rich veins of volatiles exist below its surface. The recent discovery of salt "glaciers" on the planet has provided a window into unexpectedly active geological processes. Mercury is known to hold water in its more sheltered regions, and it is thought that salts have been periodically deposited on its surface by evaporation of transient pools of water released by volcanic activity. Sometimes exposed by asteroid impacts, large masses of salt may even slowly flow under the surface. All this geological activity and chemical variety mean it is not entirely impossible that somewhere on Mercury life may cling—an astounding possibility for a planet whose daytime surface temperature is 430°C (810°F).

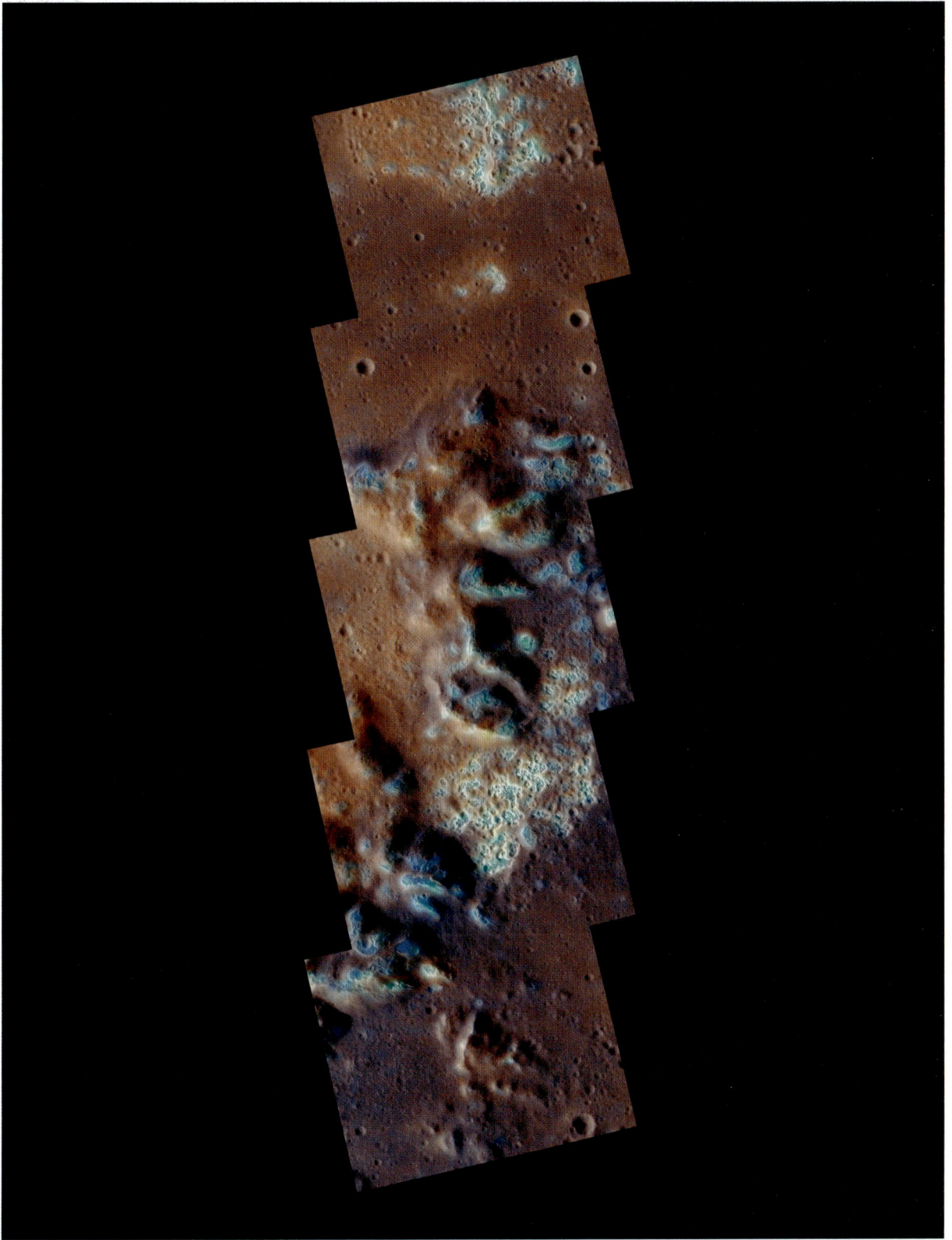

4

USE

The history of "practical" geology started long before *Homo sapiens*, but today presents us with the greatest challenges facing our ingenious species. A safe place to live, raw materials to create artifacts and buildings, and large amounts of energy—human civilization is entirely built upon geological resources. Indeed, the prehistory and history of humans are often subdivided according to how we exploited the Earth—stone, bronze, iron. This is partly because this exploitation paralleled our technological progress, but partly because its durable products outlast other vestiges of our lives.

Thai Oil refinery complex, Si Racha, Thailand, 2020.

Oil refineries, and the chemical plants often sited next to them, are among the largest manifestations of humans' exploitation of the Earth's geological resources.

"Behold, I will stand before you there on the rock in Horeb; and you shall strike the rock, and water will come out of it, that the people may drink."

Exodus 17:6

There are several key watersheds in human prehistory—bipedal walking, brain enlargement, language, agriculture, settlements, and tool use. There is evidence of hominins in Africa using stone tools more than three million years ago. No doubt stones were used initially to strike things or for throwing. The ability of our bipedal ancestors to throw stones at marauding predators must have revolutionized their lives, and may explain why wild animals find our puny species strangely threatening. Although our ancestors certainly did more with their hands than throwing and striking, it seems likely that handling stones influenced the evolution of our dexterous hands with their large opposable thumbs.

Stone tool use seems to have progressed in a particular order. It was not long before our ancestors discovered that broken stones often yield fragments with concave fracture surfaces edged with sharp, cutting ridges. As technology advanced, the deliberate, skilled crafting of stone tools became ever more complex until humans were honing double-sided, double-edged cutting and piercing implements (page 168). Later, stones for

Stonehenge, England, 2024.

Standing proud on Salisbury Plain in the county of Wiltshire, Stonehenge was built in at least three phases between 3100 and 1500 BCE. By no means the oldest stone construction in the world, the monument provides clear evidence that particular huge rocks were selected to build it—and were transported from southwest Wales, 390 kilometers (240 mi.) away, and the Marlborough Downs, 40 kilometers (25 mi.) away, using prehistoric technology.

Artist unknown, *Steinkohlenformation II* (Coal Formation II), c. 1890.

In the middle of the industrial age, this was the view of where its all-important fuel was formed—in the hot, humid forests of the carboniferous period.

toolmaking may have been the first items traded over long distances—for example, flint from the Blackdown Hills on the border between the English counties of Devon and Somerset is known to have been shipped throughout Europe.

The most remarkable, profound, and sudden transition in human history was the coming of agriculture at the end of the last ice age, around 11,500 years ago. We do not know why, but over a surprisingly short period, apparently independently in several regions of the world, a large number of people ended their hunter-gatherer existence and settled down to farm crops and animals. The effects this had on every aspect of human life cannot be overstated.

In the field behind the author's house are the eroded remains of a prehistoric hillfort, which neatly demonstrate how agricultural humans came to depend on geological resources. To farm, people must settle in one place, and it must be defensible from predators and competing tribes—and this is often dictated by topography, with high ground, islands, and river meanders being popular sites. The hillfort encloses a spring which disgorges water from between the rock strata, a reminder that wherever they live, humans must have a reliable supply of fresh water for themselves, their livestock, and their crops. Also, the fields around the hillfort yield an annual crop of fat lambs, a sign that the land possesses essential trace elements such as cobalt, copper, and selenium. Other minerals are important, too: salt (sodium chloride, $NaCl$) is also essential for humans and animals, although it was often traded rather than extracted onsite. And sometimes the high ground is not always the best—in some mountainous regions, iodine has washed from the soil, leaving children at risk of congenital hypothyroidism.

Once settled, humans could for the first time accumulate more possessions than they could carry. This included stone- and later metal-edged plowing implements, a variety of other tools, and ceramic pottery made from the nearest clay deposits. Clays are fine-grained deposits of aluminum silicate minerals—for example, $Al_2Si_2O_5(OH)_4$—which become hard when baked, and due to its durability pottery is one of the key artifacts now studied by archaeologists. Ceramic bricks and unfired clays were also used, along with hewn stone, to construct the robust buildings in which some settled humans now lived. Indeed, settlements grew to the size of modern towns remarkably soon after the coming of agriculture—"civilization" literally means "living in cities."

The earliest known geological map is part of this practical geological tradition. Dating from around 1150 BCE, the huge, colored Turin Papyrus was created for Ramesses IV's expeditions into the desert, recording the distribution of hills, gravels, mines, gold deposits, and stone quarries in a region northeast of Thebes, now Luxor.

During the several millennia after the advent of agriculture people started to purify metals. Few solid elements are found in pure "native" form in nature—gold and silver are so unreactive that they are found in metallic form uncombined with other elements, and the non-metal sulfur condenses as a pure yellow encrustation around volcanic vents. Most others must be smelted: chemically extracted from oxide and sulfide mineral ores—iron ore is a mixture of oxides, while lead ore, galena,

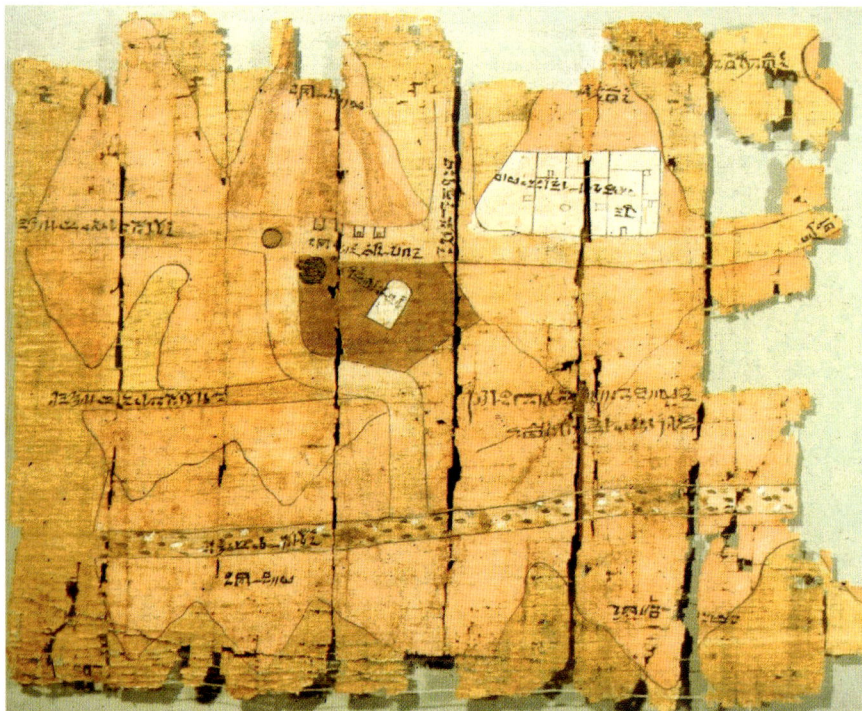

Turin Papyrus Map, Luxor, Egypt, c. 1150 BCE.

The world's first geological map, and possibly the oldest surviving folded document, the Turin Papyrus Map accurately depicts mineral resources outside ancient Thebes, now Luxor, in Egypt.

is lead sulfide (PbS). Some metals are easier to smelt than others, requiring no more than heating in the presence of air or carbon, and unsurprisingly these were the first to be extracted—metals requiring electrical separation or complex chemistry had to wait until the modern era.

Lead is an unusual example of a metal whose production may be traced through history, because it has left a trail of contamination stretching from a time four thousand years ago when it was almost non-existent in its pure metallic form, to being a ubiquitous pollutant today (page 186). Presumably at horrific cost to their slaves' health, the Romans purified large amounts of lead for their ingenious plumbing (*plumbum* is the Latin word for lead). Large ingots are still occasionally excavated where they were discarded, presumably as the empire declined. The Romans also made extensive use of concrete. Roman concrete included volcanic ash, although modern concrete is made from limestone, clay, and gypsum—and its production causes approximately 8 percent of the world's carbon dioxide emissions.

The Greek geographer Strabo (64 BCE–24 CE) wrote extensively about mining and quarrying, as later did Georg Bauer, also known as Agricola (1494–1555, page 172) in his 1556 *De Re Metallica* (On the Topic of Metals). However, extraction of metals often remained the preserve of the people who made their living from it—prospectors, miners, refiners, and assayers. What little was recorded during the medieval period suggests a thorough understanding of which rock formations contain valuable ores and where in those formations the richest ore is found. Ores were classified not only by the metals they contain, but also by how the metal was sequestered in a "combined" form—in what we would now call a chemical "compound" with other elements, such as oxygen or sulfur.

The use of geological resources to generate energy took longer. Well beyond the Middle Ages wood was the main fuel for domestic heating and cooking, and its derivative charcoal provided most carbon for smelting

Georg Bauer (Agricola) (1496–1555), illustration depicting smelting from *De Re Metallica*, 1556.

Agricola's writings give us an unparalleled insight into preindustrial metallurgy. Here an artisan is shown smelting iron in a furnace packed with burning fuel.

ores. Formed from dead vegetation, especially during the Carboniferous and Permian periods 360 to 250 million years ago, coal has been used for smelting iron in China for perhaps three thousand years, and for heating for almost as long. In Europe coal was used sporadically in areas where it lay exposed at the surface, but it was not until the thirteenth century that this source was exhausted and mining via shafts and tunnels started in earnest. Coal production soared in the eighteenth century, especially in Britain, as it formed the primary fuel for the industrial revolution (page 173).

Formed from fossilized algae and plankton, petroleum has probably been in use somewhat longer than coal, perhaps four thousand years in China, but it has only attained its dominance in the last century and a half. Crude oil may be separated into different fractions, ranging from thick bitumen, though fuel oil, diesel oil, and kerosene, to volatile gasoline, often topped by a layer of natural gas. Early uses included sealing containers and boats with bitumen, and burning kerosene for lighting. Oil was also used in a variety of ingeniously destructive medieval incendiary weapons. However, it was the invention of the internal combustion engine in the last half of the nineteenth century that led to a surge in demand for oil (page 180) and it is now also used for heating, and as the raw material for a bewildering range of chemicals.

William Williams (1727–1797), *Morning View of Coalbrookdale*, 1780.

The modern industrial age started in a few scattered locations across Great Britain, and Coalbrookdale in the Ironbridge Gorge of Shropshire, England (see page 173) was the first place that iron was smelted using coking coal.

Attempts to extinguish the burning Deepwater Horizon oil rig, 2010.

The quest for oil has led companies to seek it in ever more inaccessible locations. Floated in 2001, and capable of working at water depths of 3,000 meters (10,000 feet), in 2009 BP's Deepwater Horizon rig drilled the world's deepest oil well at 10,421 meters (34,189 feet). However, it is best known for the catastrophic explosion in 2010 which killed eleven crew and led to the pollution of large stretches of the US gulf coast.

This chapter can only skim the surface of how the human race has exploited the Earth's geological resources. If one were self-centered, one might assume that the planet had been deliberately stocked with everything we needed to build our technological civilization—fertile, safe places for our population to increase; ninety or so elements, each with its own unique uses; energy sources far beyond what human or animal exertion can provide. And I have not even mentioned tunneling, soil, fossil water, artificial fertilizers, or nuclear fuels yet.

Yet all this has come at a formidable price. Environments have been polluted and destroyed, and the climate itself is changing alarmingly. Most of what we use is not replenished at anything like the rate at which we are plundering it. And when developed countries try to switch to "clean" technologies, they discover there is still a great deal of filth and destruction involved, even if it damages landscapes and communities far away across the world. There is now a geopolitical tangle, as rich oil states desperately try to diversify their monolithic economies away from petroleum extraction, while other, poorer nations are suffering the ravages of a new era of geological colonialism.

Settling ponds at Skouriotissa copper mine, Cyprus, 2022.

In the three thousand years up to the fall of the Roman Empire, Cyprus was the world's largest producer of copper. Copper was the first metal to be smelted, and it is usually extracted from mixed copper-and-iron sulfide ores such as chalcopyrite ($CuFeS_2$). Copper was also the first metal to be deliberately alloyed with another metal, tin, to create bronze.

The copper seams on Cyprus lie in uplifted crust derived from the mid-ocean ridge (page 89) which widened the Tethys Ocean, separating the supercontinent Pangaea into its northern and southern fragments, Laurasia and Gondwanaland (page 87). Part of Tethys survives today as the Mediterranean Sea. Copper-rich lavas burst out of the mid-ocean ridge, and hot water percolated through its rocks, dissolving minerals as it went. That hot water emerged at hydrothermal vents, or "black smokers" (page 235) and the copper salts came out of solution as they mixed with cold sea water—to fall to the sea floor, and later be uplifted, exposed by erosion, and excavated by the miners of Cyprus.

Copper extraction has left many ancient slag heaps across the island, and smelting led to its dramatic deforestation. A more positive outcome of Cyprus' copper heritage is the name of both the island and the metal. The island was probably named "Kypros" after the word for the metal in the now-extinct Cypriot language; then the Romans "renamed" the orange-brown metal "cuprum" after the island whence most of it came.

Wheal Coates tin mine, near St. Agnes, England, 2009.

Tin has been produced in the southwestern English peninsular county of Cornwall for four thousand years, and for most of that time it has been traded internationally—a tin ingot found off the coast of Israel has been shown to have been made from Cornish tin around 1000 BCE. Tin is relatively uncommon, and its abundance in Cornwall may have been the only thing many Mediterranean people knew about the British Isles. Eastern Mediterranean traders described sending ships through the "Pillars of Hercules" into the Atlantic to import tin from a remote island in the ocean. In the medieval period, Cornwall was divided into administrative "stannaries'" to control tin production: the metal's Latin name is *stannum*.

Cassiterite (tin oxide, SnO_2) is often plentiful where igneous granite meets sedimentary shale, and it washes down watercourses to settle in mud, from which it may be distinguished by its dark color. It may also be mined from exposed outcrops. During the Industrial Revolution advances in explosives and water pumps allowed tin and other ores to be mined from deep in the ground—in the nineteenth century two-thirds of the world's copper was mined in Cornwall.

Stone and bronze and iron—the "three ages"

FROM 1200 BCE TO TODAY

"Stone Age," "Bronze Age," and "Iron Age" are all part of the common lexicon: pithy shorthand, evocative names with a hint of romance and even heroism, to specify three great eras of human progress according to the materials our ancestors used to make their tools.

Around 700 BCE the Greek poet Hesiod was the first to devise an implement-based chronology of human history, although his was predicated more on the idea of decline than progress—with a golden age followed by silver, bronze, "heroic," and iron ages. Today's "three-age system" was largely formulated in the 1830s by Christian Jürgensen Thomsen (1788–1865), the secretary of Denmark's Royal Commission for the Preservation of Antiquities. Thomsen was not an academic, but he made a well-intentioned attempt

to "liberate" the items in his care from what he saw as the speculative theories of archaeologists. And in his collections at least, surviving artifacts seemed to fall into three sequential eras—those made of stone, copper, and iron.

The stone age was immensely longer than the other two—it is said to have ended around 3000 BCE in the Mediterranean and Middle East. Thereafter, although metallic copper alloys are occasionally found in nature, copper and tin proved relatively easy to smelt and alloy to make bronze, although brass (copper and zinc) was also produced during the "bronze age." The bronze age is often said to have ended around 1200 BCE, in the "bronze age collapse" of many societies around the eastern Mediterranean, although some archaeologists think that collapse is somewhat overstated. Although iron is more difficult to smelt than copper, and early

The iron sword of Tamassos, near Politiko, Cyprus, c. 600–480 BCE

wrought iron was no harder than bronze, iron ore is more geographically widespread so there was less need for elaborate trading systems. In fact, according to the three-ages system, we are still in the iron age.

The three-age system has been increasingly criticized and ironically now seems a relic of the past. It places undue importance on the materials from which hard implements were made, potentially disregarding all other social and technological aspects of human life. However, these robust implements happen to be most of what remains of some ancient societies, and one can only work with what one has—and the same argument could be made against studying ancient pottery.

Even where the three-age system is accepted, it is obviously a simplification—and all the "ages" have been variously subdivided to provide greater "precision." Another criticism is that all societies around the world cannot be expected to have synchronously and abruptly transitioned from using one hard material to another. Finally, many historians and archaeologists question the validity of any attempt to divide human history into neat "chunks" of time.

One artifact, the Tamassos sword, makes clear just how complex ancient metallurgy was. Discovered in the sixth-century BCE Necropolis of Tamassos in Cyprus, it is an iron blade from an island which is literally synonymous with copper. Micro-computed tomography studies have shown how it was wrought from a single bar of iron, but that it was held in its hilt by silver-headed copper-alloy rivets, and its cross-guard was decorated by fine sheet silver and tin. The Tamassos sword clearly does not know what "age" it is from.

"The three-age system has been increasingly criticized and ironically now seems a relic of the past."

Bronze Corinthian helmet, the Peloponnese Peninsula, Greece, *c.* 500 BCE.

Characteristically bronze-age in both constituents and design, this impressive helmet demonstrates how the three-age system can mislead—it dates from a time seven hundred years after the supposed start of the iron age.

ABOVE Hindsgavl dagger, Fænø island, Denmark, c. 1800 BCE.

Found in 1876 on the small island of Fænø in Denmark, the Hindsgavl dagger is one of the finest stone tools known. Created almost four thousand years ago by simply hitting stones together, this 30-cm (12-in) blade is no more than 1 cm ($^2/_5$ inch) thick at its widest part, and has an elegant "fishtail" hilt. This period of prehistory in northern Europe is sometimes known as the "Dagger Period," and the Hindsgavl dagger features on the Danish 100-krone banknote.

RIGHT The Beijing–Hangzhou Grand Canal in Huai'an, China, 2022.

It is surprising that the Grand Canal of China is less famous than the Great Wall, considering it was built earlier, was a more complex engineering undertaking, and has had a far greater effect on the development of China. Building canals requires detailed knowledge of local geology, since they must be dug perfectly level, and the inflow and outflow of water must be carefully controlled. Largely complete by 609 CE, the Grand Canal is 1,800 km (1,100 mi.) long and runs north to south from Beijing to Zhejiang, cutting across the great rivers which flow eastwards into the Pacific. For more than a millennium it served as the empire's major trade route, supporting around a quarter of the world's population and allowing imperial control of trade and taxation. Frequently extended, it remains the longest and oldest canal in the world.

ABOVE **Shrewsbury, England, 2000.**

Shrewsbury is the county town of the English county of Shropshire, and probably started to coalesce in the eighth century, a few hundred years after the end of the Roman Empire. It has a particularly striking geographical location within a tight meander (page 115) of Britain's largest river, the Severn. Almost an island, Shrewsbury had great strategic importance as an Anglo-Saxon town ("Scrobbesburh") next to the Welsh Marches, the buffer zone against the Celts living in the mountainous west. The first Norman castle was built in 1074 to command the crucial "neck" of the meander, but the later medieval "English Bridge" to the east and "Welsh Bridge" to the west provided easier access. The modern inheritance of Shrewsbury's encircled location within its meander is a long stretch of beautiful waterfront, and a chaotic tangle of railway lines and congested roads.

RIGHT **Michelangelo Buonarroti (1475–1564),** *Awakening Slave,* c. 1503.

Marble has long been prized for its use in construction and sculpture. It is a metamorphic rock produced by the action of pressure and temperature on relatively pure carbonate rocks such as dolomite (a calcium-magnesium carbonate, $CaMg(CO_3)_2$) or calcite (a form of calcium carbonate, $CaCO_3$). When these minerals recrystallize during metamorphism, most of their sedimentary layering is lost, yielding a homogenous, dense stone, often run through by attractive veins of mineral impurities.

The world's most valued marble comes from the mountains around the town of Carrara in Tuscany, whose flanks have been stripped to reveal the dazzling white marble beneath. Many of the most beautiful sculptures of classical Rome and the Italian renaissance were cut from Carrara marble. Michaelangelo frequently traveled to Carrara and painstakingly selected the marble he would use for his sculptures. This image is of one of his many unfinished sculptures, apparently struggling to break free from the rock in which it is held captive.

Georg Bauer (Agricola) (1494–1555), illustration showing the use of a divining rod to find minerals and open-cast mining, from *De Re Metallica*, 1556.

Born in Saxony, Agricola's writings offer a unique window into the world of late-medieval mining and metallurgy. Born in Saxony, near the world's most productive mineral deposits of the era, Agricola trained as a physician in Leipzig and then traveled widely in what are now Germany and northern Italy. It was in Italy that he met the Dutch philosopher Erasmus, who encouraged him to publish his ideas.

On his return to Saxony he developed his intellectual (and financial) interests in the extraction of metals, although it was not until after his death that the seminal *De Re Metallica* was published in 1556. For his time Agricola was remarkably unconstrained by the authority of classical thinkers, and seemed intent on learning by observation. The book describes in detail how mine workers are managed, how mines are excavated, how water is pumped out and air pumped in, and the details of how ores are purified, pulverized, and eventually smelted. Agricola even discussed how ore seams might form, correctly suggesting that some are deposited in surrounding rocks by percolating water.

William Williams (1727–1797), *The Iron Bridge,* **1780.**

Although many techniques we now think of as "industrial" had been used for centuries in China, the true Industrial Revolution took place in Great Britain during the hundred years following 1760. Historians argue about the reasons why the Industrial Revolution occurred in this place at this time, but it seems to have been a remarkable coincidence of social, economic, and geological factors. Britain was a relatively politically and economically stable capitalist country with a moderate climate, excellent access to maritime trade, and the ongoing agricultural revolution meant that food production was outstripping population growth.

The Industrial Revolution involved several interlinked changes: new machines, increased use of energy, novel forms of transport, urbanization, and economic and political shifts. And all were dependent on exploiting the country's geological resources, especially its coal and iron, to build and power industrial machinery and railways. Housing, factories, and canals were constructed at a rate not seen before or since, and money poured in from mass exports. It is no coincidence that the Industrial Revolution was rooted in regions rich in coal, clay, and iron ore.

Completed in 1779, the Iron Bridge is around 48 km (30 mi.) from Shrewsbury (page 170) in the English county of Shropshire, and was the first major bridge in the world to be built of cast iron. It was constructed to facilitate transport of locally mined ore, limestone, and coal—iron was first smelted in Britain using coking coal in nearby Coalbrookdale (page 162).

The Iron Bridge inspired the worldwide use of iron in construction, and the 1797 Shrewsbury Flax Mill was the world's first metal-framed building—the forerunner of the modern skyscraper. Artists were quick to paint the new industrial landmarks rising above the English landscape, a kind of pastoral industrialism which seems charmingly unaware of the adverse effects of industrialization.

Coal miners in Hubei province, China, c. 1870.

Coal has a complex chemistry. High-quality coal such as anthracite is up to 97 percent carbon, while poor lignite may be only 25 percent. The carbon atoms in coal are arranged into large molecules based on interconnected hexagonal rings of atoms, tangled with chains of yet more carbon atoms and mixed with non-carbon impurities. When carbon burns in the oxygen in air it produces carbon dioxide:

$$C + O_2 \rightarrow CO_2$$

Combustion also releases heat energy—burning 1 kilogram (2 pounds 3 ounces) of coal releases approximately as much energy as a typical person cycling for eighty hours.

Coal of varying quality is mined around the world, either from deposits on the surface or via shafts to seams deep in the ground. Deep mining may involve leaving pillars of coal to support the ceiling of the chamber being excavated, or hydraulic jacks may be used and later removed, often leading to the collapse of the abandoned mine. Sometimes the land above may become uneven or unstable as old mine workings collapse beneath it.

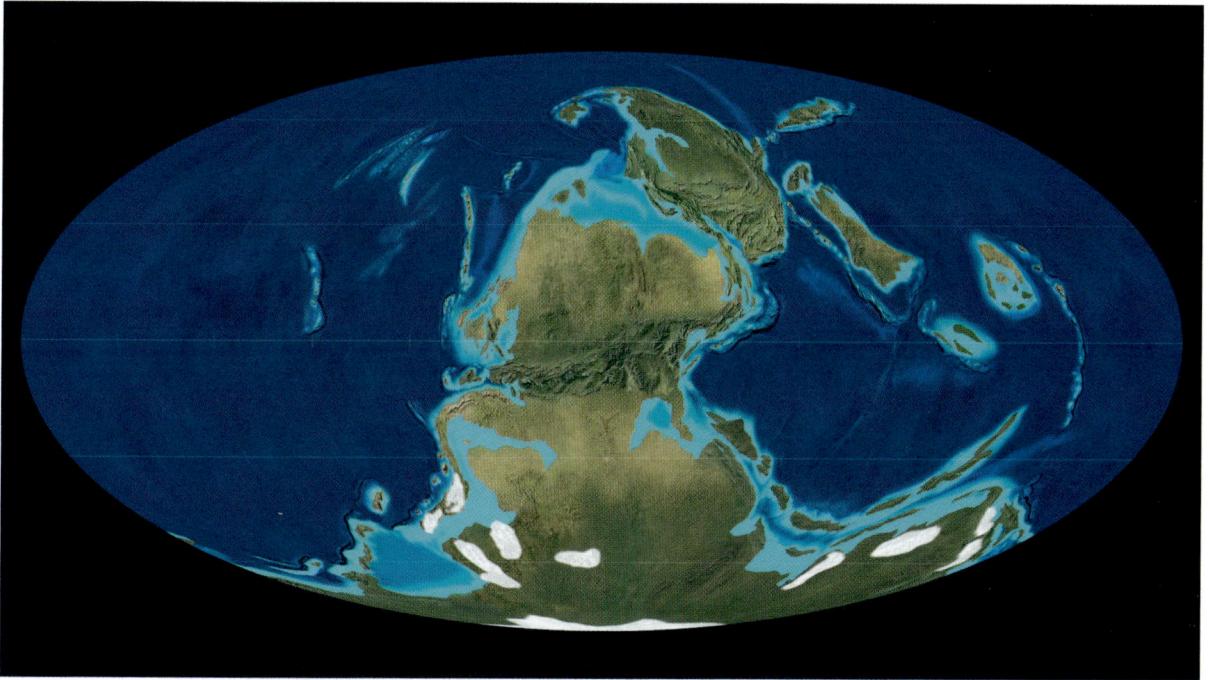

Global paleogeography of the late Carboniferous period, 300 million years ago.

Coal is the remains of ancient land plants, and has been forming ever since plants evolved to survive on land around 450 million years ago. However, most of the world's coal formed in the Carboniferous and Permian periods 360 to 250 million years ago—this era started when plants had become copious, large, and often woody, and may have ended because decomposing organisms evolved the ability to digest the structural molecules which allow land plants to stand tall. Once plants could be speedily decomposed by microbes, not much was left to form new coal.

We often think of coal as forming in hot, humid forests and swamps, and indeed this was sometimes the case. However, this was actually a time of global cooling. Water was sequestered into glaciers on land, which lowered global sea levels (page 111) to expose the continental shelves fringing the continental tectonic plates. These newly exposed continental margins became the site of large river deltas, and it is here that coal also formed.

High atmospheric levels of oxygen and carbon dioxide drove rapid plant growth, so dead plants piled up faster than they could decompose. If this material was then buried by silt, no further oxygen could reach it, and it turned to peat. Once buried deeper than 100 m (330 ft), pressure and heat starts to convert peat into coal— a process that takes millions of years, and longer for higher-grade coal.

A two-carat rectangular step-cut diamond with a garnet inclusion in the table facet.

In 1866 fifteen-year-old Erasmus Jacobs was playing on his parents' farm near the Orange River in South Africa when he found some transparent rocks. Within a few years more diamonds had been extracted in South Africa than had been found in India in the preceding two millennia. Initially diamonds were picked from riverbeds, but soon they were being mined from masses of igneous "kimberlite" that thrust through the region's ancient rocks.

Diamond is a very different form of carbon, and almost never forms from coal. Its atoms are arranged in an extremely regular transparent crystalline array with almost no space for impurities—minuscule contamination with nitrogen confers a yellow tinge, and boron a blue color. Diamond is the hardest naturally-occurring substance and it also "bends" light more than most materials, causing it to sparkle.

Diamond can form by metamorphism of crust minerals subducted deep into the Earth, or by the extreme energies released by extraterrestrial impacts, and some meteorites may themselves convey diamonds to Earth. However, most diamonds form 200 km (125 mi.) down in the mantle by the action of extreme heat and pressure on carbon—often "primordial" carbon remaining in the mantle from the planet's formation. It was once thought that the high mantle temperatures which form diamonds last existed 3,000 million years ago, and that no new diamonds had formed since then, but recent evidence suggests that some may be "only" 1,000 million years old.

Once formed, diamonds may be transported to the surface in rapid volcanic upsurges—in which "diamondiferous" magma travels upward at 40 km per hour (25 mph) to solidify near the surface as columnar, slab-like or irregular kimberlite "pipes."

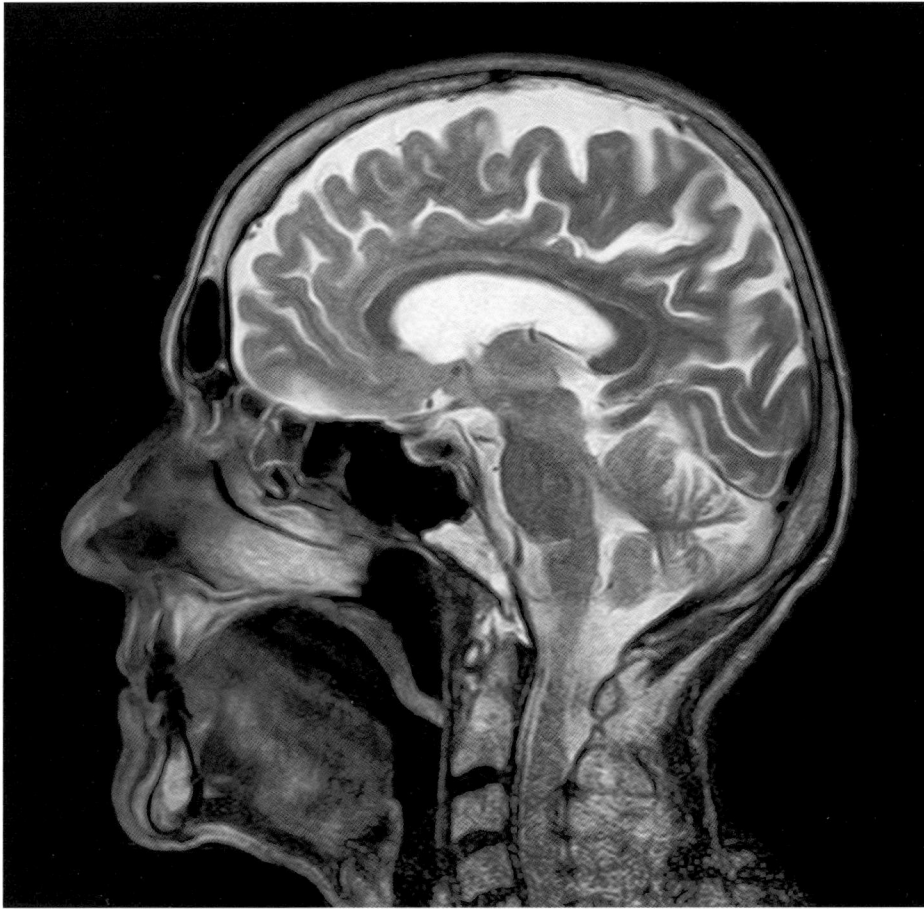

Magnetic resonance image of a midline section of a human head.

Helium (element 2) was discovered late, considering it is only the second element in the periodic table (page 39) and accounts for 24 percent of the mass of the elements in the Universe (element 1, hydrogen, makes up 75 percent). However, helium is hard to detect. It does not react with other elements under normal conditions, and there is little helium in Earth's atmosphere because this light gas escapes into space. Instead, helium was first detected indirectly in the spectrum of light radiating from the sun during the solar eclipse of 1868, and was named after Helios, the Greek god of the sun.

However, helium was soon found to be continually but slowly produced on Earth—emanating from uranium ore, for example. One of the most common products of radioactive decay is alpha particles, clusters made up of two protons and two neutrons. Alpha particles happen to be identical to the nucleus of a helium-4 atom—so all that is needed is the acquisition of two electrons to make helium gas. Most helium generated by radioactive decay remains trapped in the Earth or is lost to space, but some accumulates in pockets underground, mixed with natural gas collecting under upward geological folds ("anticlines," page 121).

The United States is the world's largest producer of this evanescent geological resource, much of it derived from the rocks around Amarillo in Texas. Helium resources are running worryingly low, yet this gas which is so keen to escape forever into space remains essential for many scientific, medical, and industrial processes. This image is a section of a human head produced using a magnetic resonance imaging machine—which requires supercooled liquid helium to allow its large magnets to function effectively.

The economics of helium are controversial—it is alarming that an essential, irreplaceable, dwindling substance is lost forever when a child's cheap birthday balloon bursts.

Center-pivot irrigation, New Mexico, 2018.

Much of the water used in the United States is fossil water, pumped from underground aquifers where it has resided for thousands of years. Twelve thousand years is sometimes used as the threshold beyond which water is considered "fossil," but some can be much older than that. Geologists can "date" underground water reserves by studying different isotopes (page 45) of hydrogen and oxygen in this ancient H_2O.

Fossil water has been exploited especially in the central and southern arid regions of the United States, and most is used for agriculture, including the center-pivot irrigation in the image. Discovered in 1898, one of the largest aquifers in the world, the Ogallala Aquifer, extends beneath parts of Colorado, Kansas, Nebraska, New Mexico, Oklahoma, Texas, and Wyoming, and may contain 10 billion tons of water.

It remains unclear whether human water extraction is depleting the aquifer faster than it can replenish. The water in the Ogallala may be fully replenished only every twenty thousand years, but there have been few surveys of how many wells are actually deep enough to tap it. However, annual snowmelt from the Rocky Mountains is the other major source of water for the arid US, and snowfall is decreasing due to climate change, so pressures on the country's aquifers will only intensify.

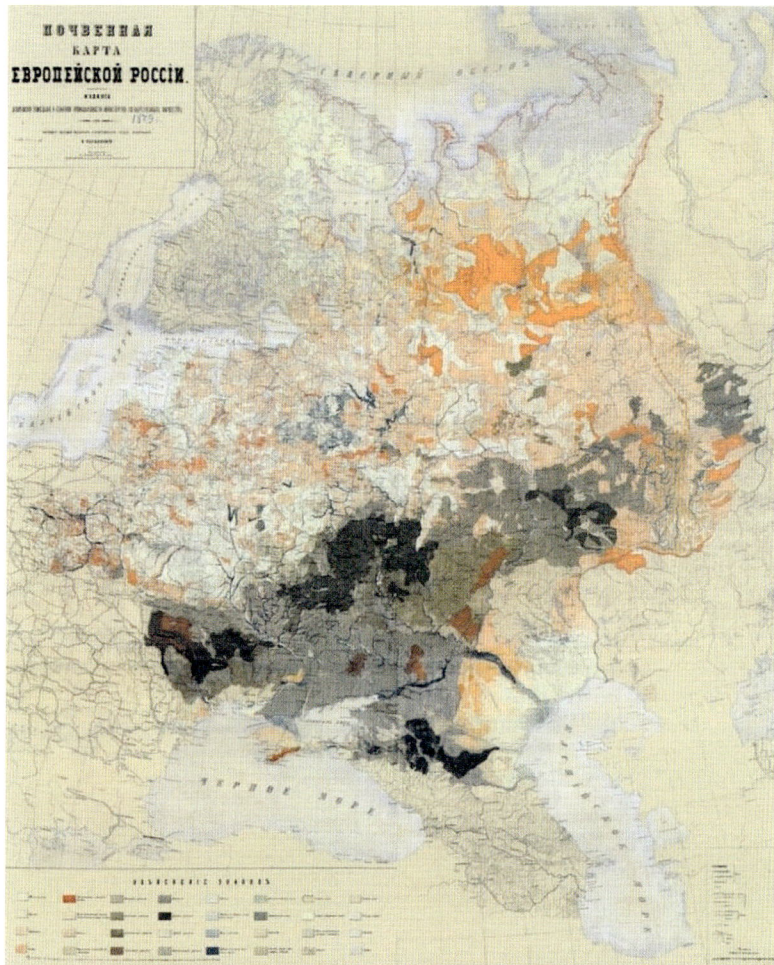

Vasily Dokuchaev (1846–1903), *Soil Map of European Russia*, **1879.**

Soil is a complex, compound material essential for human existence. It is a relatively thin layer, yet continually interacts with bedrock beneath, the atmosphere above, and the organisms that live in it and on it. It is a mixture of minerals weathered into small particles by water erosion or freeze-thaw cycles (page 142), chemicals produced by the dissolving of minerals, living organisms, dead organic matter, water, and air—very much the interface between the mineral and living worlds.

Considering soil's vital importance in agriculture, soil science developed relatively late. Professor of Mineralogy and Geology at St. Petersburg University, Vasily Dokuchaev (1846–1903) studied the layers, or "horizons," evident in soil as it transitions from mainly rocky deep down to more organic near the surface. He also investigated how soils form in different locations, depending on prevailing geology. Some of the most dramatic soil-forming processes occur on the flanks of volcanoes, as lava fragments and dissolves to form the substrate in which fertile soil can form. In Dokuchaev's native Russia, however, most soils are more mature, existing in stable equilibria with the rocks, atmosphere, and organisms around and within them.

Signal Hill oil field, Long Beach, California, 1927.

Today it seems surprising that Los Angeles' prosperity was originally based on fruit, lumber, and oil. Across the globe, surface petroleum seeps had been used for lighting and waterproofing for four millennia, but large-scale exploitation of underground reserves commenced in the middle of the nineteenth century. The Los Angeles City Oil Field started to yield large quantities in the 1890s, and the skyline of downtown L.A. was soon dominated by derricks—more than one thousand by the turn of the century. In a few decades time California produced one quarter of the world's oil. Yet oil riches brought danger to Los Angeles—a litany of oil floods, fires, and explosions, as well as the crime that wealth attracts.

Most petroleum forms from dead marine plankton and algae that settle to the seabed where they are covered by silt. Similar to coal, the first stages of oil formation involves bacterial digestion of the dead material in the absence of oxygen, in this case yielding liquid "kerogen." If kerogen becomes buried deeper than 1 km (0.6 mi.) and heated above 180°C (360°F) over several million years, it decomposes to form the hydrocarbons which constitute crude oil and natural gas. Unlike ores and coal, however, petroleum is fluid and can migrate considerable distances to a region where it eventually accumulates—often collecting under the ridge of an upward "anticline" fold due to its buoyancy.

Many petroleum hydrocarbons are chains of carbon atoms with hydrogen atoms attached, and these burn in air to yield carbon dioxide and water. For example, octane (C_8H_{18}) is the major constituent of gasoline:

$$2\ C_8H_{18} + 25\ O_2 \rightarrow 16\ CO_2 + 18\ H_2O$$

And similar to coal, combustion of hydrocarbons releases heat energy—1 liter (2¼ U.S. pt.) of oil liberates as much energy as a person cycling for eighty hours.

Fueling Mechanical War

BOMBING THE PLOIEȘTI OIL FIELD, AND LONDON PEA SOUPERS

Until the twentieth century the primary fuel for battle had been food for animals and people. However, with the advent of mechanized war, belligerents faced a new and urgent challenge—to mobilize sufficient oil to supply their war machines and transport columns. To have a chance of victory, oil must be refined and available, when and where needed, often at the end of long, vulnerable, and expensive supply lines.

During the Second World War, the Third Reich imported 60 percent of its oil from its ally Romania, especially from the Ploiești oil field. Formed at the eastern end of the Carpathian Mountains where patterns of sedimentation and sideways flow created hydrocarbon "ponds" between 3 and 5 million years ago, these oil reserves were among the largest known at the time, and had been the site of the construction of the world's first large oil refinery in 1856.

The allies were determined to cut the Reich's supply from Ploiești, and on August 1, 1943 they launched Operation Tidal Wave, sending bomber aircraft from Benghazi in Libya to destroy the oil fields' infrastructure. Yet the Germans had intercepted and decoded allied messages, and the bombers found the region encircled by artillery and

barrage balloons. Not only that, due to navigational errors, they entered the fray over a dangerously protracted period, and some were even engulfed by the explosions they themselves had caused. More flight crew were killed than died on the ground, and it was not until well into 1944 that the allies significantly interrupted the flow of oil from Romania.

The intertwined story of fossil fuels and the Second World War continued long after the war itself, and even affected the victors. Postwar Britain's finances were in a parlous state, and its government sought to balance the books by exporting high-quality coal. The British public were left to burn a low-quality domestic coal called "nutty slack," which released copious pollutants. The weather in the fateful days of early December 1952 was cold and snowy, so large amounts of nutty slack were burnt in London—and along with an atmospheric temperature inversion, this led to the worst of London's notorious smogs.

For several days, 1,000 tonnes (1,100 tons) of particulates, 800 tonnes (880 tons) of sulfuric acid, and 140 tonnes (155 tons) of hydrochloric acid were spewed daily into the layer of air which clung to London's streets. Traffic and shipping were often halted, pedestrians could not see their own feet, and estimates of the death toll range from four to twelve thousand.

Britain's worst acute pollution-related disaster, the 1952 Great Smog of London was a pivotal moment in the history of environmental legislation. The UK was the first country to bring in laws to reduce atmospheric pollution, with the 1956 and 1968 Clean Air Acts. Industries and homes were forced to change to low-smoke fuels, and eventually central heating, largely fueled by the 1980s "rush for gas."

LEFT A B-24 heavy bomber over the refinery at Ploiești, Romania, 1943.

RIGHT Guiding traffic during the London smog of 1952.

"Traffic and shipping were often halted, pedestrians couldn't see their own feet, and estimates of the death toll range from four-to twelve-thousand."

Grasberg Mine, Papua, Indonesia, 2022.

Making its first ore shipment in 1972, the Grasberg Mine in Papua, Indonesia, is one of the largest in the world—excavating into vast reserves of copper, silver, and gold. A demonstration of just how far people will go to access geological resources, it sits at 4,250 m (13,940 ft) above sea level, only 4 km (2½ mi.) from Puncak Jaya, the highest point in Oceania, and indeed the highest mountain on an island anywhere in the world.

The mine is served by its own tortuous road, aerial tramways, and an aerodrome, and its ore is pumped as a slurry through a 160-km (100-mi.) pipeline slicing through tropical forests to the nearest seaport. Like most mines Grasberg has damaged the surrounding area, far beyond the ecosystem destruction required to create and service it. Its waste "tailings" spill into nearby rivers, as does the acid runoff from material dumped on the sides of the mountain. In 1977 a rebel group attacked the slurry pipe, which led to savage reprisals by the Indonesian army in which hundreds of local people died.

Mass Rapid Transit tunnel, Singapore, 2024.

Tunneling is an essential feature of our modern world, creating underground routes for people, goods, and waste. In general, it is far more laborious and expensive than building above ground, and the methods used depend on the geology of the region. In the past, hard rock was often drilled and blasted away, although large railed modern excavating machines are now used. Alternatively, shallow tunnels may be dug as trenches which are then roofed and reinforced—this "cut and cover" technique was used extensively early in the building of the London Underground and Paris Metro train systems. This image was taken in the Singapore Mass Rapid Transit system, the first section of which was opened in 1987. Threading the network through the loose sediments underlying the city has been described as "tunneling through toothpaste."

Richard Bindler, et al., diagram showing historical lead production and the lead concentrations in lake sediments from Sweden, in "Bridging the Gap Between Ancient Metal Pollution and Contemporary Biogeochemistry," *Journal of Paleolimnology*, vol. 40, pp. 755–70, 2008.

Clair Cameron Patterson (1922–1995) grew up in a small community near Des Moines, Iowa, and has been said to have achieved three times more in his career than most great scientists.

During the Second World War Patterson was recruited into the Manhattan Project at Oak Ridge, Tennessee, where he worked on the separation of uranium isotopes used in the first atomic bombs. After the war, he applied his expertise with isotopes to rocks and meteorites and was the first to obtain an accurate estimate of the age of the Earth (4,550 million years) by measuring radioactive decay products of uranium (page 45). At the moment when he realized that he was the first and only person to know the planet's age, he recalled thinking "We did it!" He explained that the "we" referred to "the generations-old community of scientific minds."

In 1948, at the University of Chicago, Patterson started to work on lead isotopes and soon found that the modern world is so polluted with lead that it was difficult to establish what was a natural, "prehistoric" background level of lead. Lead does not usually exist in its pure metallic form in nature, but artificial smelting of lead from galena (lead sulfide, PbS) has progressively increased over the last four thousand years. This is shown in the upper trace in the graph opposite, but note that the scale is "geometric"—10^0, 10^2, 10^4, and 10^6 are, respectively, 1, 100, 10,000, and 1,000,000 tons of production per year.

In 1964, while at the California Institute of Technology, Patterson published the seminal *Contaminated and Natural Lead Environments of Man* which challenged prevailing assumptions that natural and artificial sources of environmental lead are approximately equal. He argued that 1960s Americans had blood lead concentrations ranging from "prehistoric" to "acute intoxication" and that many were subject to "severe chronic lead insult." Historical data were derived first from tree rings and ceramics, and later from lake and ocean sediments and ice cores from Greenland. In general, modern concentrations in humans, animals, the soil, and the atmosphere seem to be elevated by a factor of 100–10,000. (The lower three graphs in this image show lead concentrations in parts per million, in three Swedish lake and peat deposits laid down over the last four thousand years.)

Patterson also calculated the relative contributions to pollution of industrial uses, pesticides, paint, tetraethyl lead in gasoline, and solder in food cans. Although he faced opposition from politicians, industrialists, non-governmental organizations, and other scientists, his outspoken advocacy during the 1970s and 1980s was crucial in turning the tide against the human race's rush to poison itself with lead.

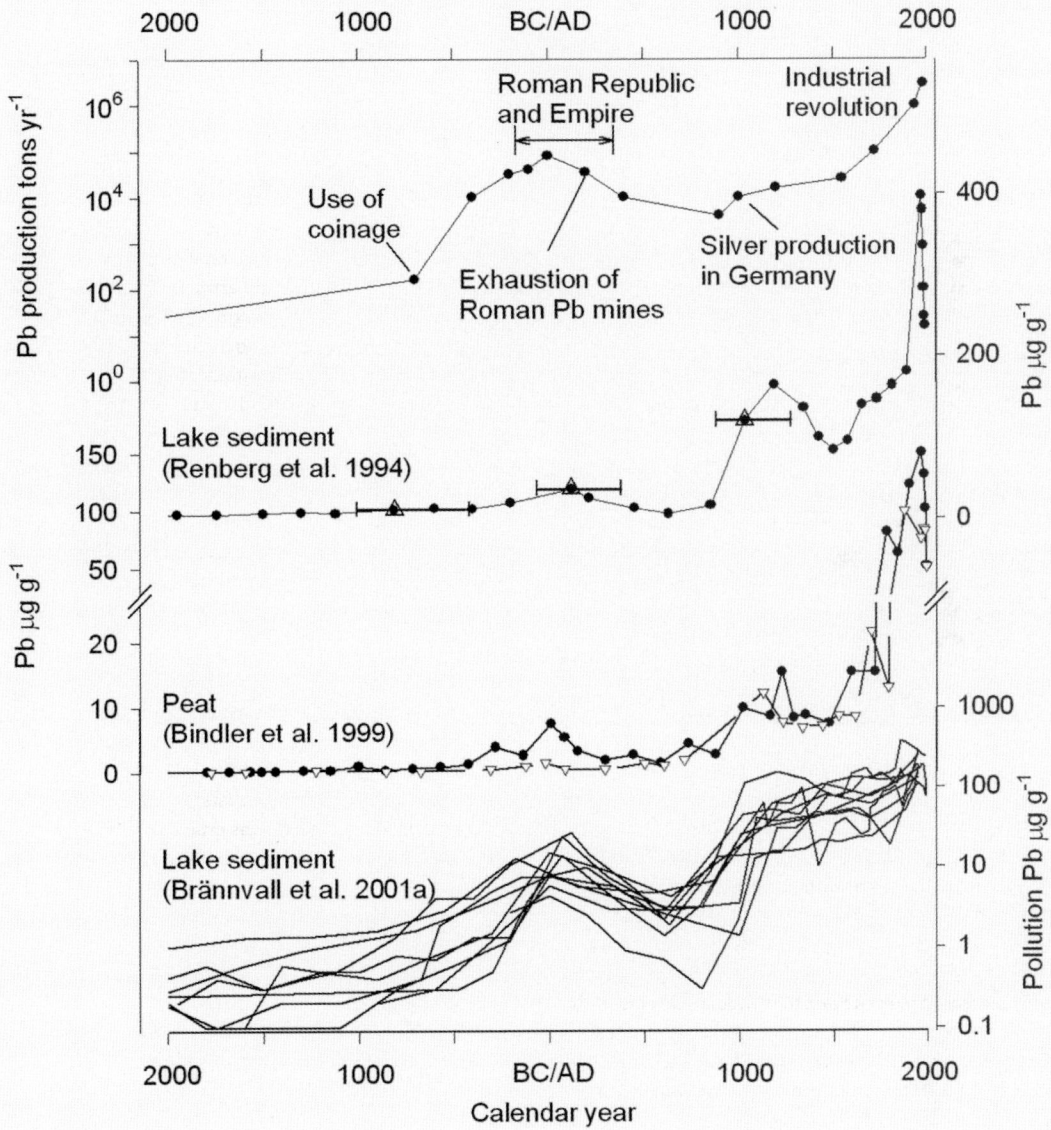

Slag heaps at La Oraya mines, Peru, 2023.

With the development of modern electronics and "green" technologies, the previously obscure forty-ninth element, indium (see periodic table on page 39) is now in demand. Indium compounds are used in solar panels, semiconductors, liquid crystal displays, and, since 1984, touchscreens. The global reserves of this and many other elements are unclear, and indium may also be extracted from the tailings dumped when more "mainstream" metals were mined in past decades. However, there has been controversy about where some of these newly valuable materials are sourced—in 2024 the Democratic Republic of the Congo filed legal complaints against Apple Inc. for claimed use of "conflict minerals" in its products.

Lithium "fields" in the Atacama Desert, Chile, 2023.

Metals toward the left-hand side of the periodic table are not usually thought of as being derived from "ores." Instead of being smelted from oxides or sulfides, their compounds are often soluble and may be mined directly as "salts" or extracted by dissolving them in water—examples include rock salt (sodium chloride, NaCl), fluorspar (calcium fluoride, CaF_2), gypsum (calcium sulfate, $CaSO_4$), and potash (various potassium salts, often used in fertilizers).

In recent years, attention has turned to the lightest metal in the periodic table: the third element, lithium. The first lithium batteries were sold in 1991 and lithium is now a key constituent of batteries used in electric vehicles. The world's largest lithium reserves are in Chile and Australia, and demand for the metal may even rejuvenate the ancient Cornish mining industry (page 165). Lithium may be isolated from spodumene ore—lithium aluminum silicate, $LiAl(SiO_3)_2$—or by pumping water into lithium-bearing rocks to dissolve its salts.

However, the lithium market is a volatile one, with variations in global production and unpredictable public adoption of electric cars causing periodic crashes and mass redundancies of mining staff. Also, there has been geopolitical concern about the domination of "green" mining in some countries by China—for example Chile (see image), Argentina, and Bolivia.

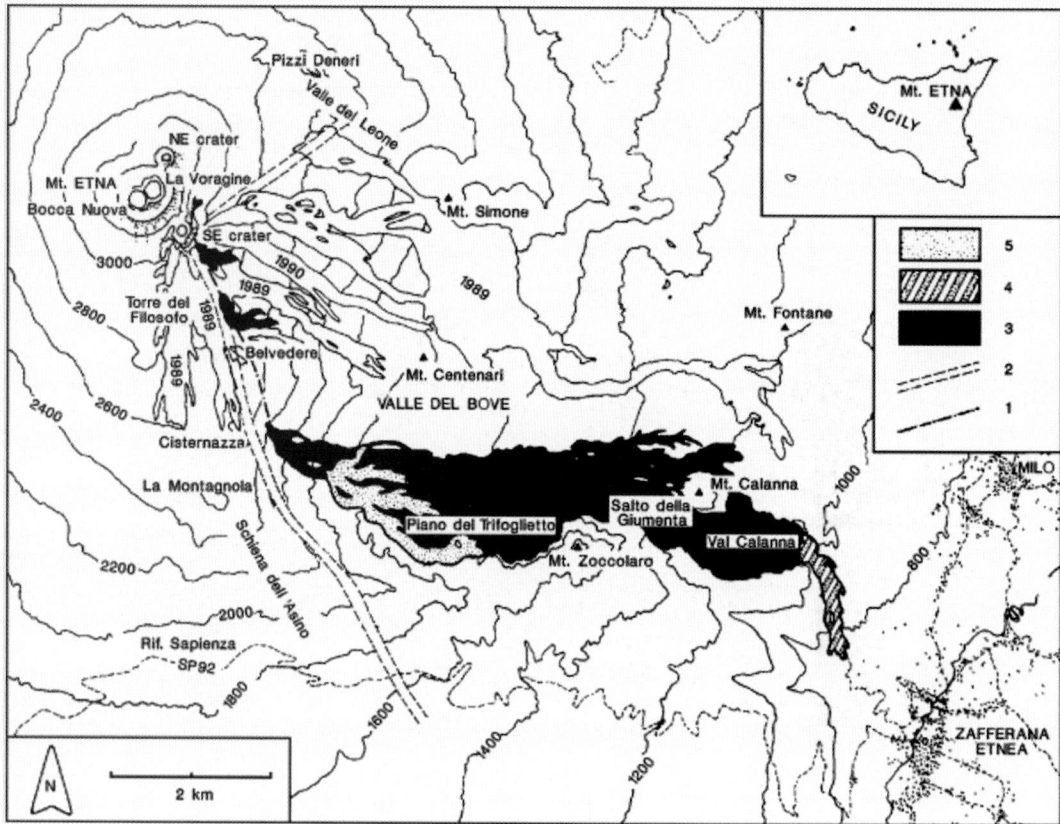

Lindsay McClelland, sketch map of lava fields, eruptive fissures, and fracture systems and lava flows from "Report on Etna (Italy)," *Bulletin of the Global Volcanism Network*, vol. 17, 1992.

Attempts to control volcanic eruptions have met with limited success, but the 1991–93 eruption of Etna on the Italian island of Sicily was a well-managed, if nerve-racking, exception. By January 1992 extensive lava flows had spread down the eastern flanks of the volcano until they threatened the town of Zafferana Etnea (main flow large black region on map, summit at top left, town lower right).

Initially the largest flow was halted 2 km (1¼ mi.) upslope from the town by a vast artificial earth dam, which held back a large lava lake until April when it was overtopped. Once that happened, lava began to flow rapidly down a narrow valley toward the town (hatched region), slowed by the construction of three smaller earthen dams. Attention then turned to reducing the flow from higher on the mountain—boulders were dropped by helicopter into the active vent, and a large hole was blasted in the side of the main lava tube, causing lava to spill harmlessly sideways onto an area high above the town (dotted region).

Shinjuku skyline, Tokyo, Japan, 2009.

With strong regulation and a great deal of money it is possible to make buildings surprisingly earthquake-proof—indeed, the main risk to human life is often the less-controllable tsunamis which follow those earthquakes. Traditional buildings in Japan have a long history of seismic resilience—small buildings were often built of bamboo, light wood, and paper to reduce the dangers of collapse, and larger pagoda-style buildings with a central flexible wood pillar resisted resonant oscillations during quakes.

Shinjuku, the "Manhattanized" area of Tokyo, has been a testbed for innovations in quake resistance. Buildings are often based on flexible steel frames rather than crumbly concrete pillars, and walls often taper toward the top and are divided into sections by flexible joints. Rather than being set into concrete, pipework and wiring run through voids to avoid rupture and fires, and doors are designed to fall away during earthquakes, so they do not block escape routes. Some buildings have rubber foundation mounts and dampers installed throughout their structures. A few even include massive rooftop pendulums to moderate "flailing" as the structure is pulled from side to side. A new innovation for smaller buildings is to install powerful air pumps which can temporarily levitate a building above its foundations to "disconnect" it from movements of the ground.

ABOVE Burning oil fields, Kuwait, 1991.

The early 1990s were not a good time for petrochemical pollution. In 1991, as Saddam Hussein's forces retreated from Kuwait at the end of the Gulf War they deliberately set around seven hundred oil wells ablaze. For most of a well's productive life, petroleum emerges under pressure, so a spewing burning well can be extremely difficult to extinguish and seal. Paul "Red" Adair became particularly famous at this time, "capping" many Kuwaiti wells so that all were sealed by November 6 of that year. Nonetheless, the region became and remains an environmental disaster zone—the smoke plume extended to 1,300 km (800 mi.), a 15-km (9-mi.) slick spilled into the Persian Gulf, and three hundred oil lakes formed. The skies are now clear, but oil has contaminated deep into the ground.

RIGHT The oil tanker *Braer* founders, Shetland Islands, Scotland, 1993.

A disaster with a less negative outcome was the wrecking in high winds of the Liberian-registered oil tanker *Braer* in Quendale bay in the Scottish Shetland Islands on January 5, 1993. Caused mainly by the captain's poor seamanship and the subsequent failure to reboard the vessel when an opportunity arose to haul it away from the rocks, the grounding released 84,700 tonnes (635,000 U.S. barrels) of light crude into the churning waves. However, although the volume of oil was four times that spilled by the environmentally catastrophic *Exxon Valdez* disaster, the *Braer* was, in many ways, a narrow escape. The heavy seas swept the light oil away from the land and fragmented it into tiny droplets, which were then digested by naturally occurring marine bacteria. There are many species of bacteria adapted to digesting oil seeping from the seabed, but most are present in low numbers—ready to proliferate when the next spill comes.

The rewards of instability

FERTILE ASH AND DESERT OASES

When one hears of a disastrous earthquake or volcanic eruption, the initial reaction can be bewilderment as to why anyone would decide to live near active volcanoes or faults. Admittedly, sometimes disasters strike where no natural disaster has taken place within living memory, but a large percentage of the world's population live within earshot of rumbling reminders of the forces waiting to be unleashed.

For most of human history, the main attraction of volcanoes has been the soils which form from volcanic lava and ash. Lava weathers and crumbles remarkably quickly and contributes elements such as magnesium and potassium to soils—magnesium is essential for the synthesis of chlorophyll, while potassium is needed for a wide range of biochemical processes. Ash has the additional advantage of already being physically fragmented so its constituents can rapidly leach into soil. Unlike lava, it may land on existing soils without burning them. Many farmers look forward to their soil being "topped up" with nutrients from the next eruption.

Over the last few thousand years, people have also exploited the mineral wealth associated with volcanic regions. Hot water coursing through rocks can deposit seams

The volcano Popocatépetl over the town of Puebla, Mexico, 2016.

Near Mexico City, the largest city in North America, the active volcano Popocatépetl looms over the lives of around twenty million people. Considered a sacred mountain for hundreds of years, its ashy eruptions have enriched the soil ever since humans first arrived in the region.

of metal ores such as copper (page 164) and ore may also be mined directly from masses of solidified magma. More recently, volcanic regions have become even more attractive as sources of geothermal energy or tourism income—and Iceland is perhaps the best example of these.

Choosing to live on seismic faults may seem even more inexplicable, but usually there are good reasons for this, too. Faults are often located near the coast and produce landforms which humans find useful—for example, San Francisco Bay (page 77) is one of the best harbors in the world.

However, one tragic example demonstrates how the effects of faults on water supply have often tempted humans to live near them. On December 26, 2003, a powerful earthquake occurred directly underneath the spectacular ancient desert city of Bam, on the southern edge of the high Iranian plateau. The quake was shallow, with most energy released at depths of only 2–8 km (1¼–5 mi.) along a sideways-moving "strike-slip" fault (page 136). The maximum sideways displacement was 2.7 m (9 ft). At least 31,000 people perished—perhaps a quarter of the city's population.

Yet it was precisely this precarious geology that had caused Bam to be founded here centuries earlier. The displaced fault has raised the water table, leading to surface drainage of water unusual for such an arid region. Throughout history, these water sources have been exploited by the building of qanats—underground water supply tunnels which spurred the development of many settlements across the Middle East. Thus Bam is an oasis in the middle of the desert, and ancient trade routes detoured through it so travelers could quench their thirst and water their beasts.

Coseismic interferogram of the Bam earthquake from a descending orbit, 2003.

Superimposed on a satellite image of the region around the Iranian city of Bam, the colors in this "coseismic interferogram" depict the surface deformation caused by the 2003 earthquake. The city itself is in the mottled region to the left of the pink-purple ovoid just above the center of the image.

Entrance to the Cheyenne Mountain Complex, Colorado, 1984.

Geological expertise is important in identifying sites where things may be "hidden away," in some cases to protect them from the dangers of the outside world. The 2-hectare (5-acre) Cheyenne Mountain Complex in Colorado is an underground bunker beneath 700 m (2,300 ft) of a mountain made almost entirely of granite. Started in 1961, it housed the North American Air Defense Command and Aerospace Defense Command, and since 2019 has housed the U.S. Space Force. A 1965 geological survey identified its "deep crustal rigidity" and suggested that it would be an excellent site for activities under "buttoned up" status—meaning it could for some days continue surveillance of incoming aircraft and intercontinental ballistic missiles, even when hit by a thirty-megaton nuclear strike, an electromagnetic pulse, or the annihilation of nearby Colorado Springs. Approximately two-thirds of a million tonnes/tons of granite were excavated in the complex's construction and its caverns now house several three-story buildings mounted on huge impact-absorbing springs.

Onkalo spent fuel repository, Finland, 2023.

Sometimes geologically "secluded" places are needed to protect the world outside from the noxious substances within, and one example of this is the nuclear waste disposal facility at Onkalo in Finland. Nuclear power is likely to be crucial in reducing man-made carbon emissions, but it generates radioactive waste, some of which will be hazardous for millennia. Buried 450 m (1,480 ft) down, near three nuclear reactors on an island in the southwest of Finland, and opening in the mid-2020s, Onkalo is the world's first "geological disposal facility." Here high-level waste radioactive over geological timescales may be stored in a geological environment known to be stable over those timescales—up to 100,000 years.

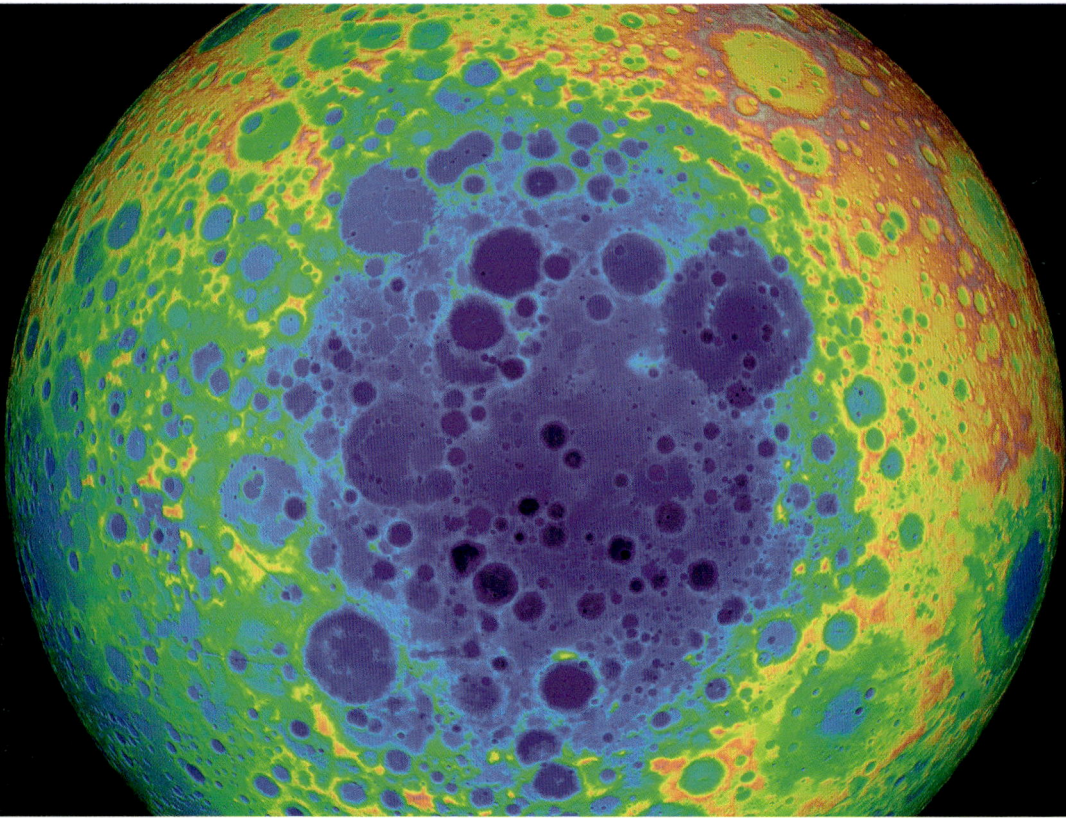

The lunar South Pole–Aitken Basin region, 2017.

Exploiting the geological resources of other bodies in the solar system is becoming a more realistic proposition—and the hope is that some can be used *in situ* on the body on which they are acquired. Crewed interplanetary spaceflight is an immense practical and financial challenge, partly because of the costs of lifting people, vehicles, and supplies from Earth to beyond Earth orbit. One solution is to manufacture supplies, and even machine and vehicle parts, from minerals available on the surface of a less gravitationally demanding world, such as the Moon.

Encircling the Moon's south pole is the Aitken Basin, at 2,500 km (1,550 mi.) wide and 12 km (7½ mi.) deep, one of the largest impact craters in the solar system. Evidence from orbiting probes, artificial impactors, and China's 2018 Chang'e 4 lander suggests that our airless satellite is the location of a remarkable 600 million tonnes/tons of water ice, and possibly ten times more. The Moon's rotation has been very stable for some time, and some sites near its poles have been permanently in shadow at -170°C (-270°F) for the last two billion years. As a result, water ice from meteorite impactors has accumulated in these frigid pockets, and water is the key to many aspects of human spaceflight—drinking, washing, and fuel production, for example. Chinese scientists are even investigating the possibility of using local minerals and water to manufacture bricks for construction on the Moon. Using lunar water sounds outlandish until one considers that lifting 1 kg (2¼ lb.) of water to the Moon costs perhaps $20,000.

Peter Rubin, artist's impression of the asteroid Psyche, 2020.

An alternative approach to off-world mineral exploitation is to mine material and then return it to Earth. The ideal candidate for mining is a body rich in valuable minerals, which comes reasonably close to Earth, and escape from whose gravitational field requires little energy—and it is for these reasons that asteroids are the most likely target.

M-type asteroids such as 16 Psyche (3D model shown in image) may be 30–60 percent metal, much of it in a native, uncombined form, and including some rare elements such as iridium. Considering Psyche is around 240 km (150 mi.) across, the potential financial gains from what are now called "easily retrievable objects" are immense. It has been estimated that mining the ten most cost-efficient asteroids would yield a profit of 1.5 trillion dollars. Psyche has been claimed to contain more than 100 quintillion dollars' worth of gold at current prices, although exploiting it would of course cause the cost of gold to collapse. It is no surprise then, that NASA is sending its *Psyche* spacecraft to orbit and map the asteroid from 2029 onwards.

Other options are available, too—C-type asteroids contain large amounts of water and require less energy to reach than the lunar poles, while the Moon may itself prove to be a viable source of the valuable isotope helium-3, which is extremely rare on Earth.

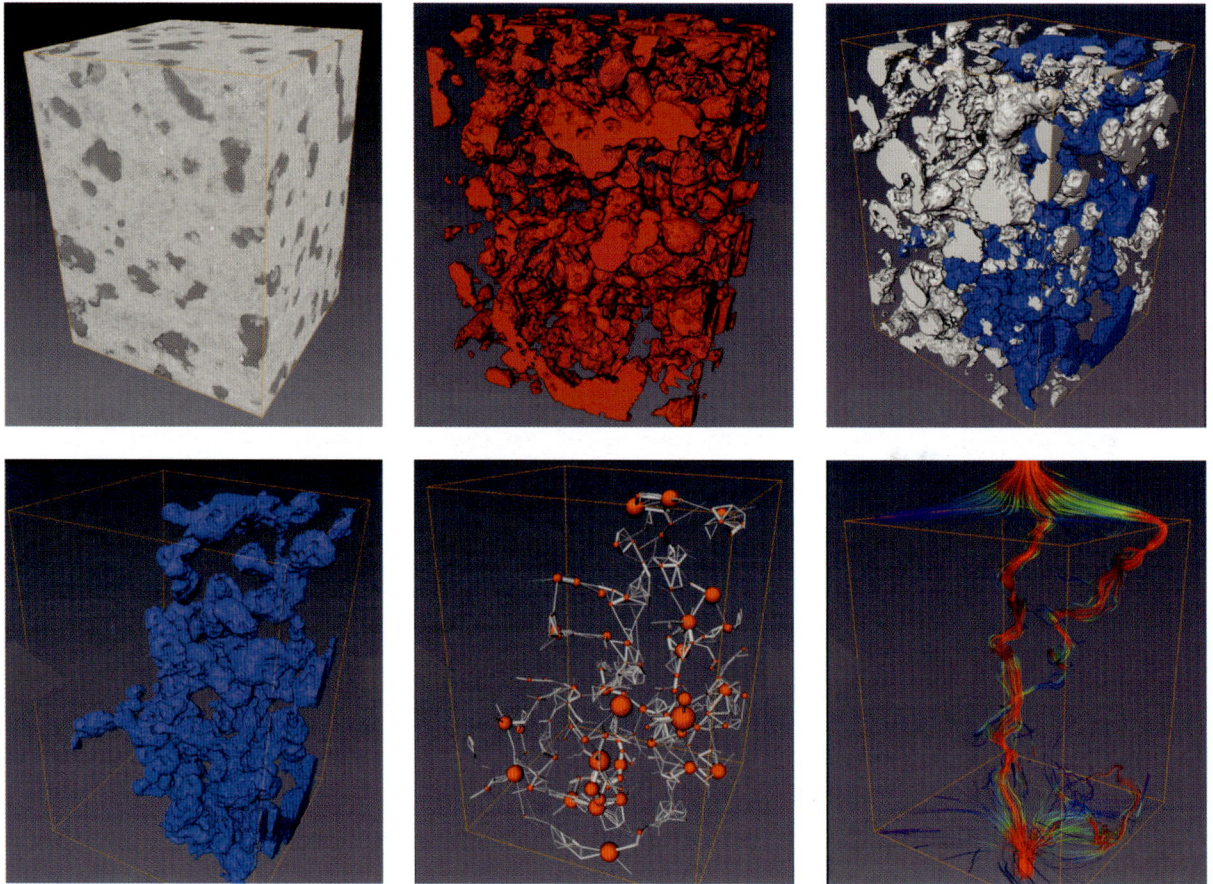

Callow, B., et al., illustration from "Assessing the carbon sequestration potential of basalt using X-ray micro-CT and rock mechanics," *International Journal of Greenhouse Gas Control,* vol. 70, pp. 146–56, 2018.

In 1856, Eunice Newton Foote was the first American woman to publish a paper on physics in a scientific journal—"Circumstances Affecting the Heat of the Sun's Rays"—in which she was the first to suggest that increasing concentrations of carbon dioxide in the atmosphere would lead to global warming. Now, 170 years later, humans are desperately trying to reduce carbon dioxide emissions and mitigate their effects on climate.

One strategy is to pump CO_2 into geological formations so it is sequestered from the atmosphere over very long timescales. This diagram is from a scientific paper investigating how basalt may be assessed for its capacity to store carbon in this way. From upper-left onwards are shown a space-filling

three-dimension model of a region of basalt, and progressive analyses (red, grey/blue) of the pores in the rock, eventually focusing on the pores which are interconnected and thus available for gas storage (lower left). The data were then processed to yield the lower middle image – a simplified map of pores (red balls) and their interconnections (grey lines). Finally, the lower right image is a model of the potential flow velocity of gas through the rock.

Porosity and mechanical properties indicated that one pilot site may be able to store around a third of a gigatonne of carbon dioxide (a gigatonne is one billion tonnes, and metric tonnes and imperial tons are similar in size). For comparison, it is thought that human activity has released 3,500 gigatonnes of CO_2 in total, 1,000 gigatonnes of which remains in the atmosphere. It has been calculated that the Earth's exposed basalt has the capacity to store up to 120,000 gigatonnes.

Victoria Fulfer, microplastics found in sediment near Providence, Rhode Island, 2023.

As well as exploiting geological resources, humans' activities are themselves altering the geology of the planet. Indeed, geological "signatures" may one day be used to determine whether other worlds have ever supported their own technological civilizations far in their distant past (page 231).

This image is of microplastics in sediments deposited in Narragansett Bay over the last twenty years. Microplastic particles have been found in myriad samples from across the world, even seeping into sediments deposited before the large-scale production of environmentally persistent plastics. From 1950 onward an exponential increase was detected, as global annual plastic production increased from 2 million to its current level of 400 million tonnes/tons.

5

LIFE

Geology and the study of life are entwined disciplines, and they have at times seemed indistinguishable: the rocks tell us the context of the creatures, and the creatures tell us the context of the rocks. Also, Earth is currently the only place in the Universe where life is known to exist, and we are still learning how geological processes helped life start, and still support it now. As a result, the old science of paleontology, "the study of ancient beings," has now been joined by abiogenesis, "origins from the lifeless."

The dinosaur *Tyrannosaurus rex*, 2017.

Discovered in the late nineteenth century in the United States, *Tyrannosaurus* remains the archetypal predatory dinosaur of the public imagination. This species existed between 73 and 66 million years ago.

> "No animal is a stone; certain animals are stone; therefore certain animals are stone and are not stone."

Ramon Llull, c. 1280

People have been aware of fossils for a long time, although they did not always know what they were. In Japan, fossil shark teeth were seen as evidence of a legendary "heavenly dog," while dinosaur bones may have inspired North American plains tribes' stories of the thunderbird. The ancient Greeks collected fossils which sowed the seeds of legends of giants and the cyclops. The earliest surviving depiction of a fossil is a giant skull painted on a ceramic bowl recounting the story of the "Ketos Troias," or "Trojan Sea Monster." And already in 500 BCE, Xenophanes (c. 570–c. 478 BCE) had suggested that fossils are not supernatural but are the remains of living beings, and may provide evidence that regions of dry land were once submerged under the sea.

Although the idea of fossilization did not fit well with Christian teachings about the age of the Earth, many thinkers argued that fossils are indeed the remains of ancient organisms. Leonardo da Vinci (see page 23) sketched animal fossils he found high in the Italian hills, and he was clear they

Corinthian bowl depicting Herakles and Hesione fighting a sea monster, the Ketos Troias, c. 550 BCE.

The "monster" in this image is thought to be based on the skull of a fossil whale, and as such may represent one of the oldest depictions of a fossil.

Conrad Gessner (1516–1565), illustration of mollusk shells from *De Rerum Fossilium*, 1565.

Gessner was one of the first scientists to compare living and fossil forms, and to suggest that ancient creatures may once have lived far from the regions where similar animals live today.

were not just *lusi naturae*—"games of nature" resulting from the Earth's playful tendency to create lifeforms from stone. Soon afterward, the Swiss philosopher Conrad Gessner (1516–1565) published his 1565 *De Rerum Fossilium* (On Fossil Objects), which illustrated the similarities between fossil and living crabs and sea urchins, and he speculated about why these aquatic species' remains are now often found far inland.

In his 1667 *Natural History of Oxford-Shire*, Robert Plot (1640–1696) was the first to depict a dinosaur bone in the context of natural history, although he claimed the bulbous lower end of a *Megalosaurus*' thighbone actually belonged to an ancient giant or Roman war elephant. Due to the shape of the bone, this dinosaur species almost ended up with the scientific name *Scrotum humanum*.

With characteristic insight, Niels Stenson, or Steno (see page 24) argued that the *glossopetrae* or "tongue stones" once believed to have fallen from the sky are almost identical to the teeth of today's sharks. He also speculated about how dead animals might fossilize, proposing the "corpuscles" which constitute living tissue are replaced by mineral "corpuscles" to create a solid, stony fossil. Steno's combined fascination with both sedimentation and fossils was the direct forerunner of the pivotal work of the English surveyor William Smith (see pages 30 and 216–217) who proposed a meticulously researched system of "stratigraphy" in which each rock layer contains the fossilized remains of its own characteristic, defining fauna. As a result, Smith's publication of his *Strata Identified by Organised Fossils* in 1816 represents the point at which geology and paleontology were most conjoined.

The early years of the nineteenth century were also witness to the two events which spawned modern paleontology and evolutionary biology. The first was the discovery in Lyme Regis, England, by twelve-year-old Mary Anning (1799–1847) of a fossil of the ancient 5.2-m (17-ft) dolphin-shaped marine reptile now called *Ichthyosaurus*. Although not the first outsized ancient creature to be unearthed, this one find was so striking, complete, and alien, that it changed the course of science—demonstrating that previously unknown extinct creatures lie locked in the Earth's rocks.

The second pivotal event was the publication in 1809 of *Philosophie zoologique* by a French zoologist at the Muséum national d'Histoire naturelle

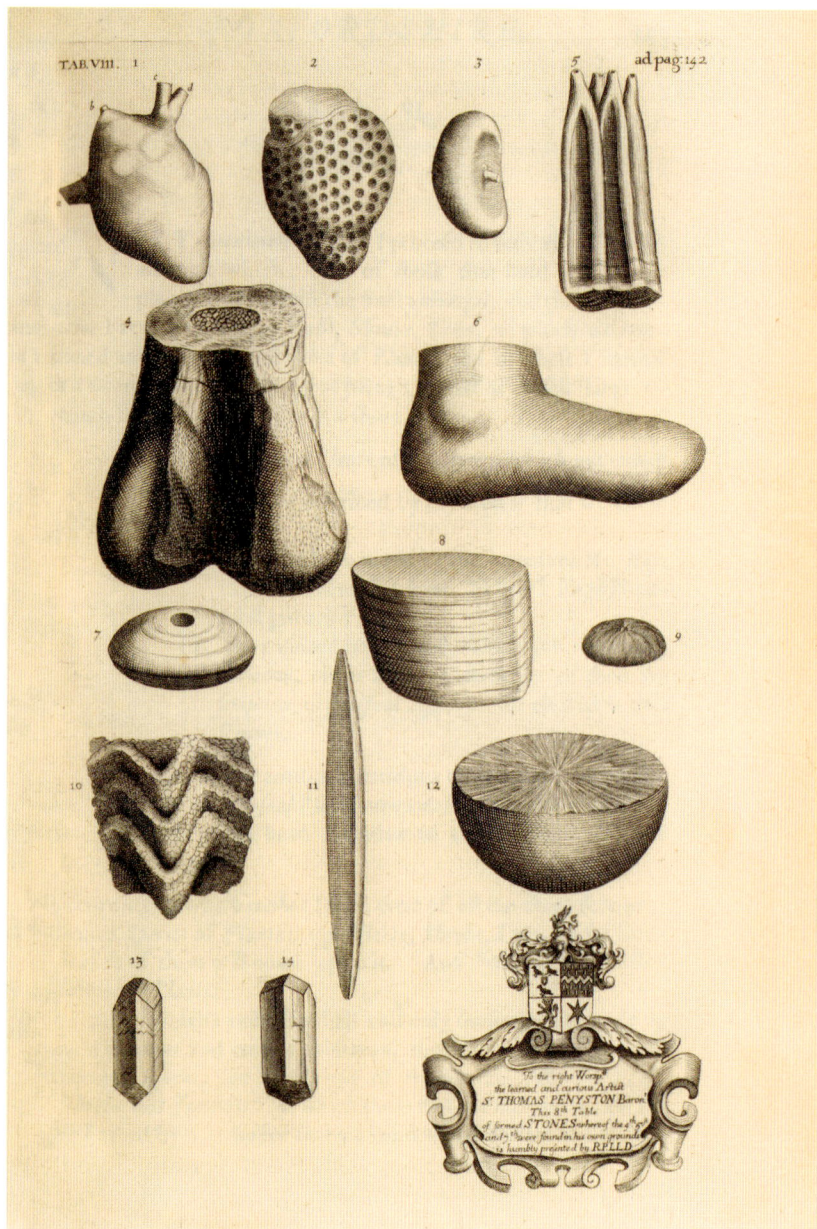

Robert Plot (1640–1696), "Eighth table of formed stones found in his own grounds and humbly presented," in *Natural History of Oxford-shire*, **1676.**

The fourth illustration in this montage was not part of the skeleton of a mythic creature or giant as Plot claimed, but rather the lower thighbone of the dinosaur *Megalosaurus*.

in Paris, Jean Baptiste Lamarck (1744–1829). Although evolutionary theories had been discussed before, Lamarck developed the idea that animal species change over time, divide into multiple "descendant" species, and become extinct—and that the diversity of life today is the outcome of these processes. He even drew some of the first evolutionary trees.

The grand unifying theory of the history of life, natural selection, was to arrive in 1858—the *mechanism* which underlies the *process* of evolution. Under the guidance of Charles Lyell, two little-known scientists, Charles

Carl Eduard von Eichwald (1795–1876), illustration showing the tree of animal life, from *Zoologia Specialis*, 1829.

Pre-dating the theory of natural selection by thirty years, this is the first diagram to depict the branching pattern of animal evolution as a tree. Between 1809 and 1859 many scientists believed that evolution occurs, but at that time no mechanism could yet explain *how* it occurs.

Darwin and Alfred Russel Wallace (1823–1913) co-published *On the Tendency of Species to form Varieties; and on the Perpetuation of Varieties and Species by Natural Means of Selection*. Darwin's ideas had crystallized while he was naturalist on the famous research voyage of HMS *Beagle* between 1831 and 1836, and he spent much of his time fossil hunting and pondering the power of geological change and extinction. Indeed, in some ways he arrived at the theory of natural selection as a geologist, inspired by Lyell and Adam Sedgwick (page 127). In contrast, Wallace's route to natural selection was as a zoologist who had focused on the geographical distribution of living species. It is intriguing that the 1858 paper represented something of a crossing-over point for the two men—geological Darwin was to spend much of the rest of his life pondering the mechanisms by which organisms evolve, whereas zoological Wallace worked on how the planet affects the species which inhabit it, the effects of glaciation on life, how human activities might damage the Earth, and even the evidence for life on Mars.

And so the scene was set for a busy century of excavation, paleontology, and evolutionary biology. Between *On the Tendency of Species* in 1858 and the coming of plate tectonics and molecular biology around the 1960s thousands of fossil organisms were discovered, characterized, and their evolutionary relationships clarified. Although there was considerable bias toward certain fossil beds in particular countries, the history of life was largely elucidated during this time, or at least the most recent 600 million years of it—when organisms were sufficiently large or robust to leave identifiable fossils. Yet throughout that time and even today, the geology-zoology dichotomy remains. Just as Darwin and Wallace brought their different outlooks to bear upon the mystery of evolution, scientists often approach paleontology with one of two fundamentally different philosophies—at heart, some want to know about animals and their history, and others want to know about the Earth's processes.

What might be called the "geological approach" to paleontology has unearthed subtle truths about the world, which are in some ways more intellectually satisfying than the headline-grabbing discovery of a ferocious-looking dinosaur. How do geological processes affect the story of life, especially now we know the continents are episodically separating, migrating, and recombining? If extinction is important in driving large-scale evolutionary change, have extinction events been caused by geological processes? How has life left its mark on geology—are the planet's rocks changed by the living things which inhabit its surface? How reliable is the fossil record, why are there "gaps" in it, and can we expect to find all the "missing links" in the evolutionary story? And are the fossils we unearth representative of the past communities of organisms, or do we only see a selective, skewed subset of what was there?

Fittingly, one of today's major biological questions links directly to geology—how did life on Earth start? Unless living organisms first came to Earth from elsewhere in the universe—a possibility called "panspermia"—then life must have arisen *here* from non-biological material. This "abiogenesis" is fiendishly difficult to fathom. All life is based on cells: membranous sacs containing salty water, huge DNA molecules which encode genetic information, and a bewildering array of large molecules. These large molecules do many things: they facilitate cellular processes including the acquisition and utilization of nutrients and energy, they copy the DNA when a cell needs to reproduce, and they decode its genetic code to make yet more large complex molecules. In short, the whole system is so complicated and self-interdependent that it is hard to imagine how it ever arose.

Yet arise it did, and apparently quite quickly. The Earth is 4,540 million years old (page 18–19) and it is possible that life was present remarkably soon after the planet's surface became a moderately stable environment—maybe 4,000 million years ago. Whenever it happened, life began when the Earth was very different from now—its internal processes, its chemistry, its atmosphere, its water. In 1952, Stanley Miller (1930–2007) and Harold Urey (1893–1981) conducted the most famous experiment in the science of abiogenesis when they applied heat and electrical discharges to a mixture of water and the gases thought to dominate the primordial atmosphere. Nothing crawled out of the flask, but amino acids were created and these are the building blocks of many life molecules.

The Earth's minerals provide a far more diverse set of ingredients than the Miller-Urey experiment did, and their crystalline structures have even been suggested to have acted as scaffolding on which early biomolecules coalesced. We now know that many celestial bodies contain large amounts of liquid water, but the most likely location for extraterrestrial life is where that water is in contact with rocks (page 247).

"Unless living organisms first came to Earth from elsewhere in the universe—a possibility called 'panspermia'—then life must have arisen *here* from non-biological material."

Stanley Miller working in the laboratory, 1953.

This is the lab where the Miller-Urey experiment took place, in which an attempted recreation of the conditions and chemistry of the early Earth yielded some of the basic constituents of the chemistry of life.

However, while life on Earth may have started due to its mineral chemistry, this does not mean life is an inevitable result of geology. Probably other conditions must be met. We know our planet lies in the sun's "habitable zone" where solar radiation engenders a surface temperature conducive to life based on liquid water. However, we do not know whether life started on the surface, or deep in the sea, or even within rocks, and we also do not know if abiogenesis was powered by energy sources other than sunlight. The Earth also has other unusual features which may have been important for the appearance of life—a magnetic field that deflects harmful cosmic rays (page 104), a disproportionately large satellite which stabilizes its rotation (page 48), and plate tectonics which continually refresh its surface (page 90).

These geological features of the Earth are unusual, exceptional, and some are unique—for now, at least. If they were as crucial for the origins of life as some believe, then life on our planet may be a rare and precious thing indeed.

本草品彙精要卷之二十三

獸部上品

鱗蟲

龍骨
無毒　白龍骨齒角
吉弔紫稍花等附

骨龍

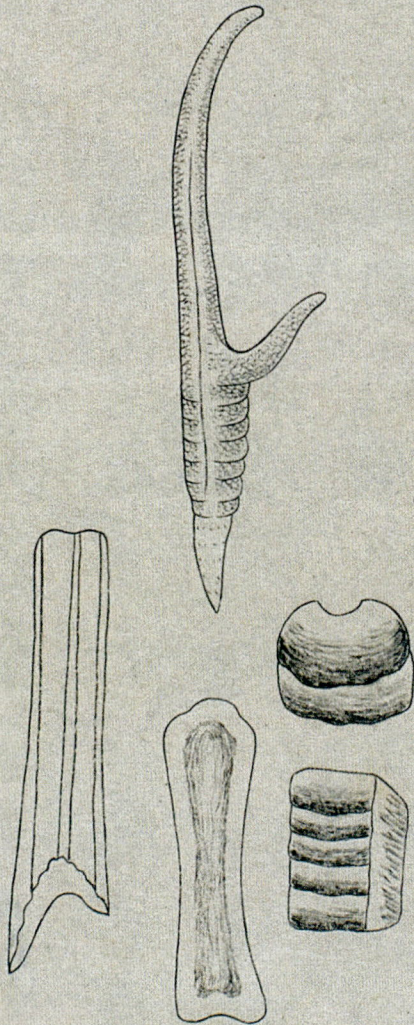

LEFT Liu Wentai et al., illustration of a dragon bone, from *Bencao pinhui jingyao* (Materia Medica), 1505.

In China fossil bones were often called "dragon bones" and were sometimes even ground up for use in traditional medicines. Dragon bone was claimed to reduce anxiety and stop diarrhea. This image is from a forty-two-volume herbal compiled on the orders of the Ming Emperor Li Zong. Ironically, that same year the emperor became feverish and died, because the author was unable to cure him.

ABOVE Johann Jakob Scheuchzer (1672–1733), *Homo diluvii testis* described in *Lithographia Helvetica*, 1726.

The Swiss biologist Johann Scheuchzer firmly believed that fossils are the remains of ancient organisms, but also that they became embedded when the Earth's rocks recrystallized after the Biblical flood. *Homo diluvii testis* means "Man, Witness of the Flood" since Scheuchzer thought this fossil was the remains of an ancient sinner staring out at him from the rock. In the nineteenth century the specimen was correctly identified as a giant salamander.

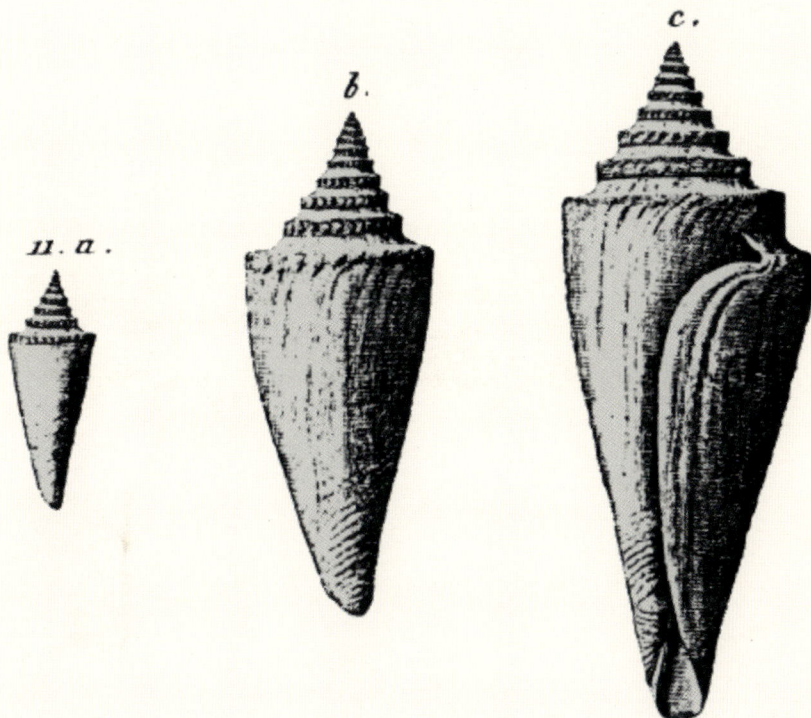

LEFT Robert Hooke (1635–1703), "Snake stones," *Posthumous Works*, 1705.

It was the English enlightenment polymath Robert Hooke who did most to establish that fossils are the remains of creatures which died in the ancient past. As a child he was an enthusiastic fossil hunter on his native Isle of Wight, and he later became an excellent draughtsman—he drew many of his beautiful scientific illustrations himself. As a scientist he had a strong belief in the value of unbiased observation, and described in detail the intricacies of fossil animals and plants as if they were alive today. In his *Lectures and Discourses of Earthquakes and Subterraneous Eruptions* he asserted "There have been many other species of creatures in former ages, of which we can find none at present."

"Snake stones" are actually ammonites—mollusks related to cuttlefish, squids, and octopuses—which thrived in the sea from approximately 400 million years ago until the late-Cretaceous extinction event (page 236) 66 million years ago. Ammonites were abundant, varied, and underwent rapid evolutionary change, making them the perfect fossil timepieces for dating rock strata.

ABOVE Giovanni Battista Brocchi (1772–1826), illustration of *Conus antidiluvianus*, from *Conchiologia fossile subapennina* (Subapennine Fossil Conchology), 1814.

Trained in law and theology at Padua in northern Italy, Giovanni Battista Brocchi roamed freely between different disciplines in the arts, humanities, and sciences. He noticed that styles of art change over the decades, often being replaced by a succession of new artistic movements, and he believed he could see that same process taking place in the fossil record. He often studied mollusks, whose shells of calcite and aragonite—two forms of calcium carbonate ($CaCO_3$) which fossilize readily—show trends, extinctions, and replacement as one scans through rocky strata. Brocchi even wondered if each animal type can last only for a certain pre-ordained span. Indeed, Brocchi's thinking paralleled that of two of his contemporaries: Lamarck's ideas about evolution (page 206) and Smith's new system of stratigraphy (pages 216–217).

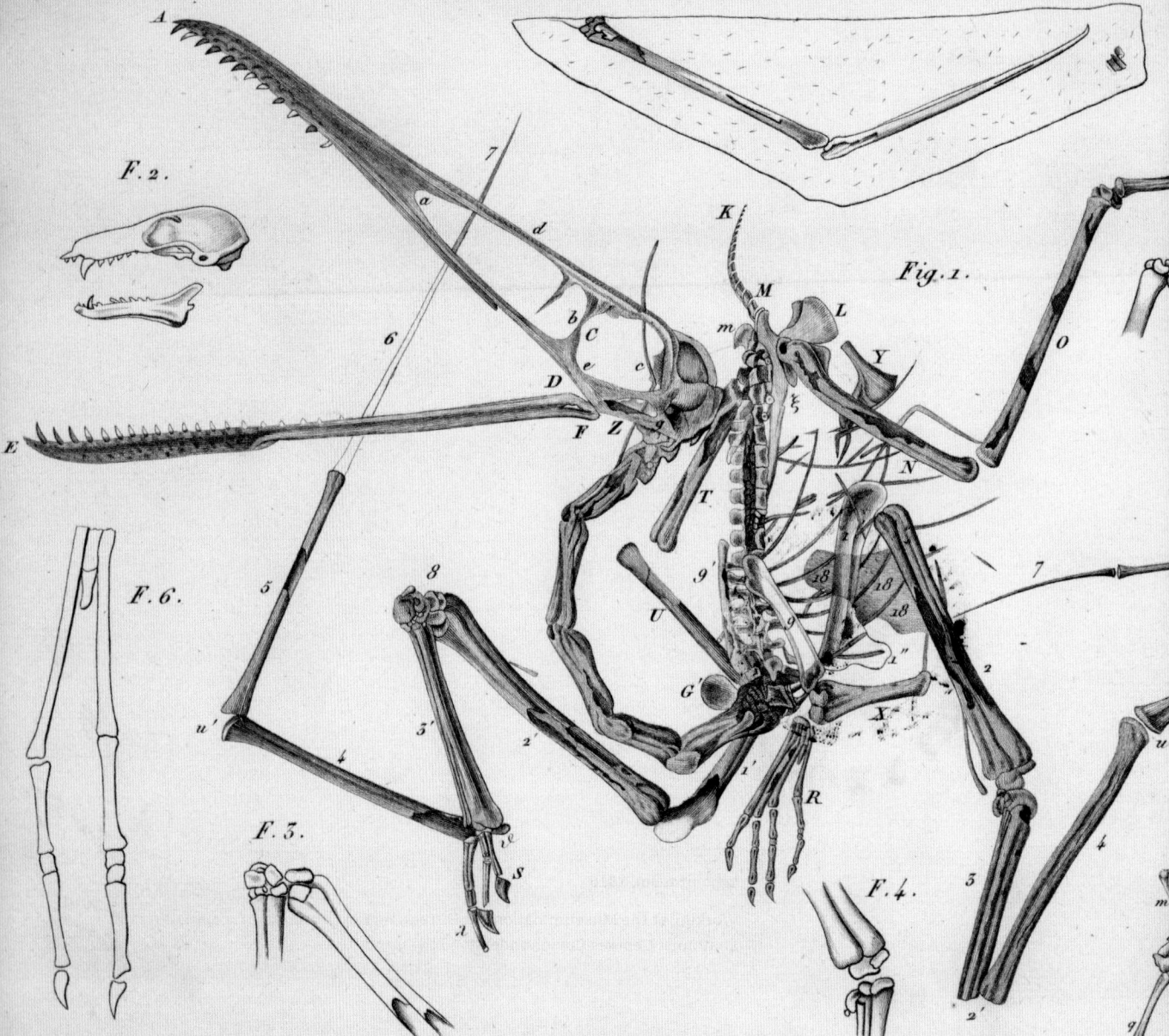

PTÉRODACTYLE.

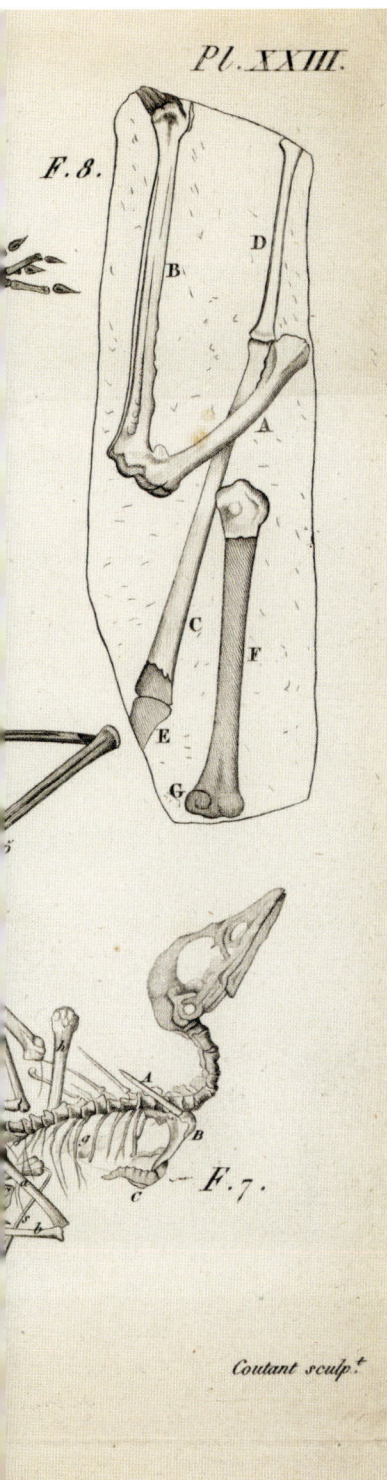

ABOVE Georges Cuvier (1769–1832) and Alexandre Brongniart (1770–1847), illustration of the rock strata of the Paris Basin, from *Essai sur la géographie minéralogique des environs de Paris* (Essay on the Mineralogical Geography of the Environs of Paris), 1808.

LEFT Georges Cuvier (1769–1832), illustration of a pterodactyl, from *Le regne animal: distribue d'après son organisation (The Animal Kingdom, Arranged According to Its Organization)*, 1816.

Working at the Muséum national d'Histoire naturelle, the French comparative anatomist Georges Cuvier understood the importance of geological context to the new science of paleontology. Cuvier was famed for touring Europe's fossil collections and identified many major extinct groups of backboned animals ("vertebrates")— including pterosaurs (such as the "reptile volant," *ptérodactyle*), many marine reptiles, giant ground sloths, and a variety of elephant-like creatures. It was also Cuvier who correctly identified Plot's *Scrotum humanum* as the thighbone of a huge extinct reptile (page 206) and Scheuchzer's *Homo diluvii testis* as a giant salamander (page 211).

In the first decade of the nineteenth century he joined forces with the geologist Alexandre Brongniart (1770–1847) to survey the strata of the rocks in the Paris Basin—analyzing a stratigraphic column of approximately 300 m (1000 ft) containing mollusk-shell-rich sediments deposited when the region was a shallow sea during the Jurassic and Cretaceous periods. Despite his unparalleled contribution to the study of extinct animals, Cuvier was surprisingly disparaging of the nascent theory of evolution.

Organized fossils and faunal succession

WILLIAM "STRATA" SMITH (1769–1839)

Born into an Oxfordshire farming family, the English surveyor William Smith was to travel the length and breadth of Great Britain cataloguing the surprising geological variation present in that small and temperate island. Geological surveying was enormously important in late-eighteenth- and early-nineteenth-century Britain as mines, canals, and railways spread across the nation. And when Smith inspected his mines and cuttings he noticed something which struck him as a sign of a wider truth: wherever he traveled in the island, the layers or rock always seemed to be arranged in the same order.

As "Strata" Smith's nickname makes clear, he was fascinated by the regular layering of sedimentary rocks, and he published the first detailed modern geological map (page 31). In addition, he emphasized that each stratum contains distinctive fossils, which can be used to define that stratum just as clearly as its position or constituent minerals. He called this correspondence between rocks of different ages and animals of different types his "Principle of Faunal Succession"—a bold

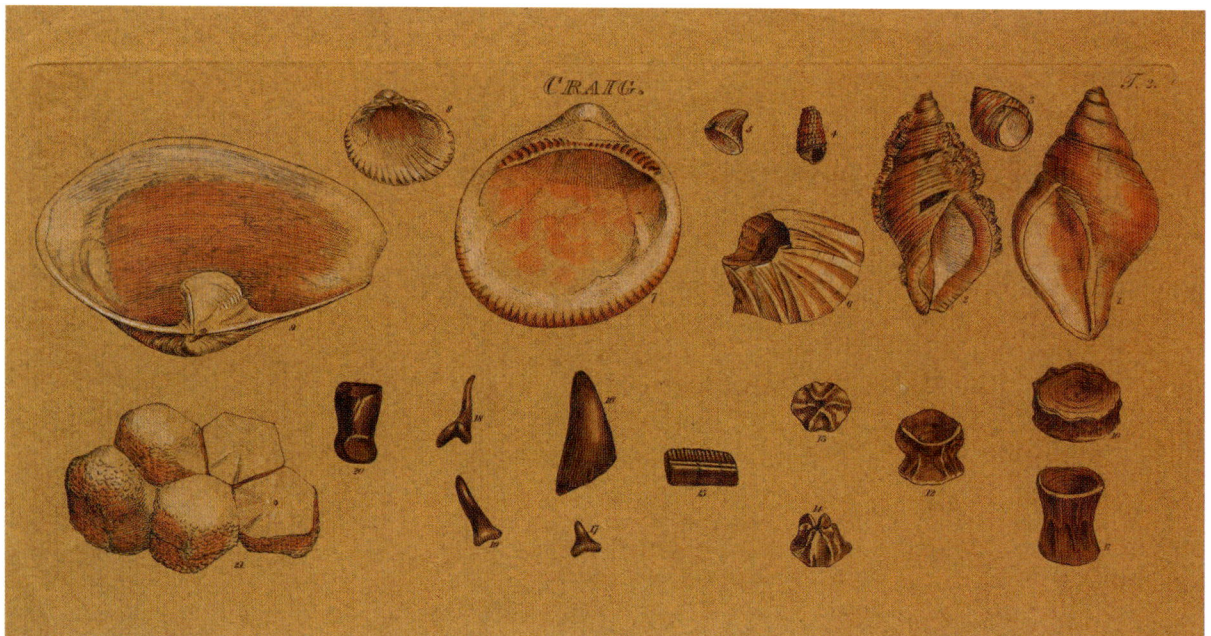

CRAIG.

name chosen to emphasize how geology and paleontology could be combined to establish the chronology of the past.

Just as important as Smith's giant map was his less florid "1817 Geological Table of British Organized Fossils," a clear expression of how the history of the Earth is set out in a vertical stack of strata beneath our feet—and color-coded to refer to his beautiful individual colored engravings of the fossil forms which characterize each stratum. By matching epochs in the planet's past to layers in its sedimentary rocks, Smith was that rare thinker who really did change how we understand the world: the rocks are immensely old, yet their history is ordered and comprehensible and parallels the history of life's long tenure.

Smith's ideas about the fossil record foreshadow many later thinkers. As one can imagine, he was particularly perplexed by what we would now call "gaps in the fossil record"—where his neat system suddenly appeared to "omit" a chunk of time. He speculated that these could be "real" gaps that occur when sedimentation temporarily ceases, or that they could be the record of episodes of mass extinction when a set of his beloved fossils vanished suddenly from the rocks, to be replaced by entirely new creatures.

Smith was not a famous man, and certainly not a rich one, so his elegant theories took quite some time to spread throughout the scientific world. Sometimes the credit for his work was claimed by others, and he was to die penniless. Only later was his contribution to the synthesis of geology, biology, and chronology to be rediscovered.

William Smith (1769–1839), "Craig" (opposite), "Clay Over the Upper Uolith" (right), illustrations from *Geological Table of British Organized Fossils*, 1817.

Two montages of fossils, printed on paper color-coded to correspond to entries in Smith's *Geological Table*. According to his "Principle of Faunal Succession," each layer of sediment contains its own distinctive fossil fauna—so fossils may be used to establish the chronology of rocks, and rocks used to establish the chronology of fossils.

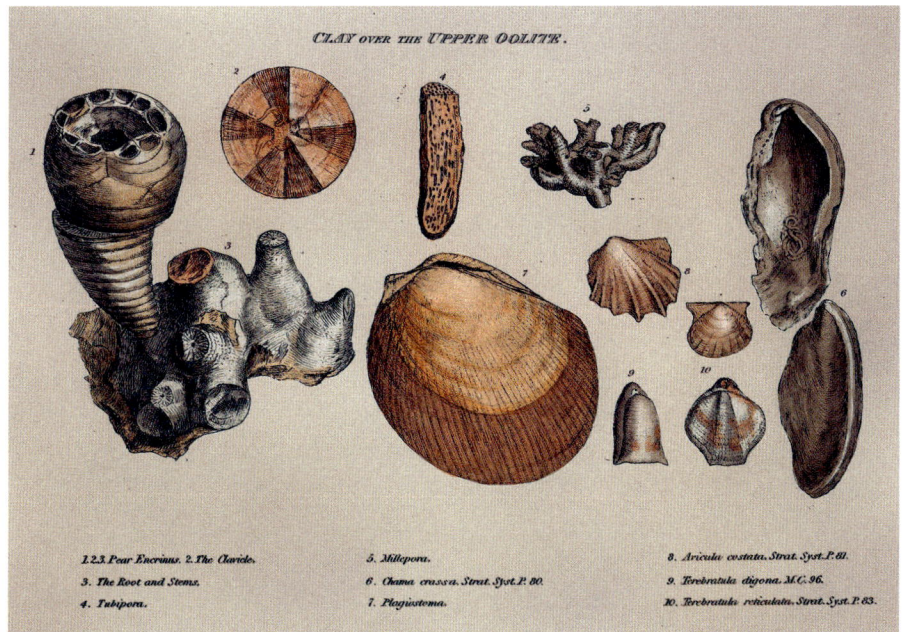

CLAY OVER THE UPPER OOLITE.

1.2.3. *Pear Encrinus*. 2. *The Clavicle*.
3. *The Root and Stems*.
4. *Tubipora*.

5. *Millepora*.
6. *Chama crassa. Strat. Syst. P. 80.*
7. *Plagiostoma*.

8. *Avicula costata. Strat. Syst. P. 81.*
9. *Terebratula digona. M.C. 96.*
10. *Terebratula reticulata. Strat. Syst. P. 83.*

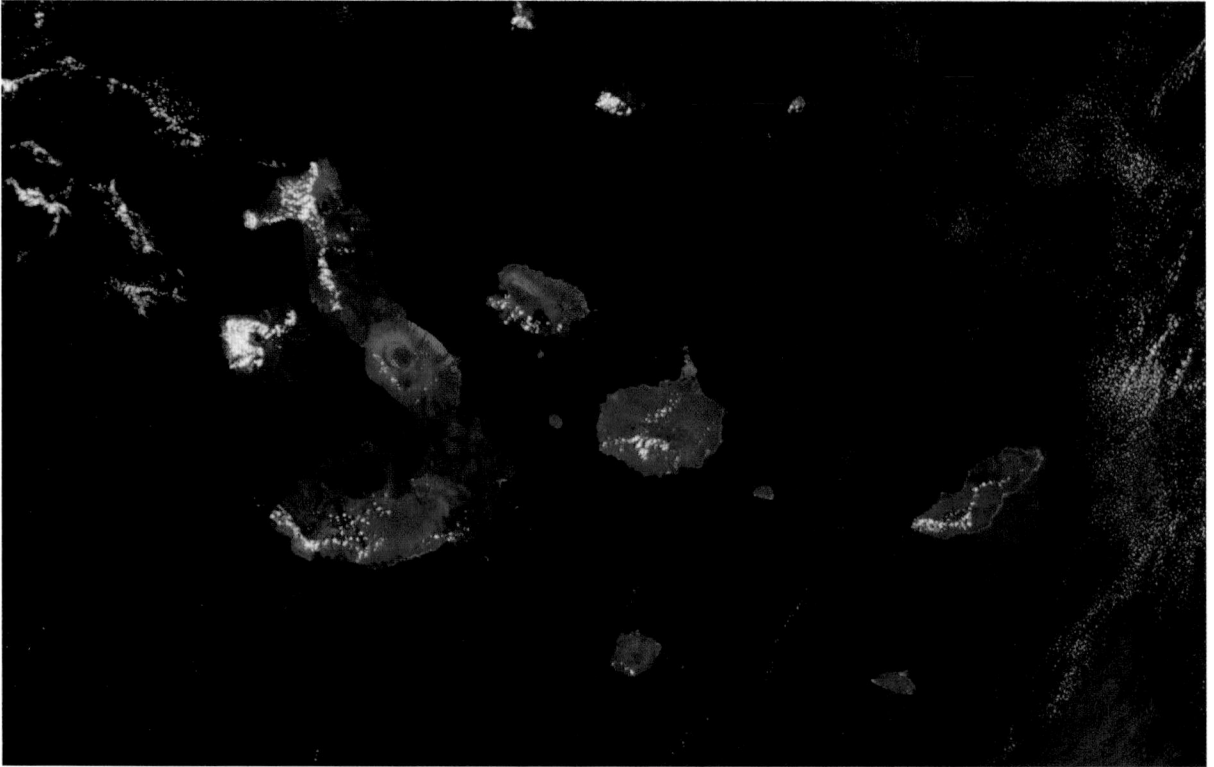

Terra satellite image of the Galapagos Islands, 2002.

Charles Darwin's famous contributions to science were steeped in geology—
some of his observations during the voyage of HMS *Beagle* focused on the interface
between biology and geology. Also, sometimes geological phenomena created
unusual "natural experiments" for him to study.

Some of the most famous discoveries which contributed to Darwin's theory of
natural selection were made on the Galapagos Islands—an archipelago of more
than one hundred volcanic islands 1,000 km (600 mi.) west of the continent of
South America. Like Hawaii and Yellowstone (page 102), the Galapagos are being
continually formed by eruptions taking place over a "hotspot" deep in the Earth's
mantle. And like Hawaii, the tectonic plate onto which the islands are erupting
is slowly moving sideways—to the east.

Because of this movement the newest islands are found in the west, forming on
"pristine" ocean crust. Straggling toward the east the islands become older and
more eroded—up to four million years older. Thus, the Galapagos present a range
of island environments slowly transitioning from one side of the archipelago
to the other: an excellent place to study how animals adapt as different
opportunities present themselves.

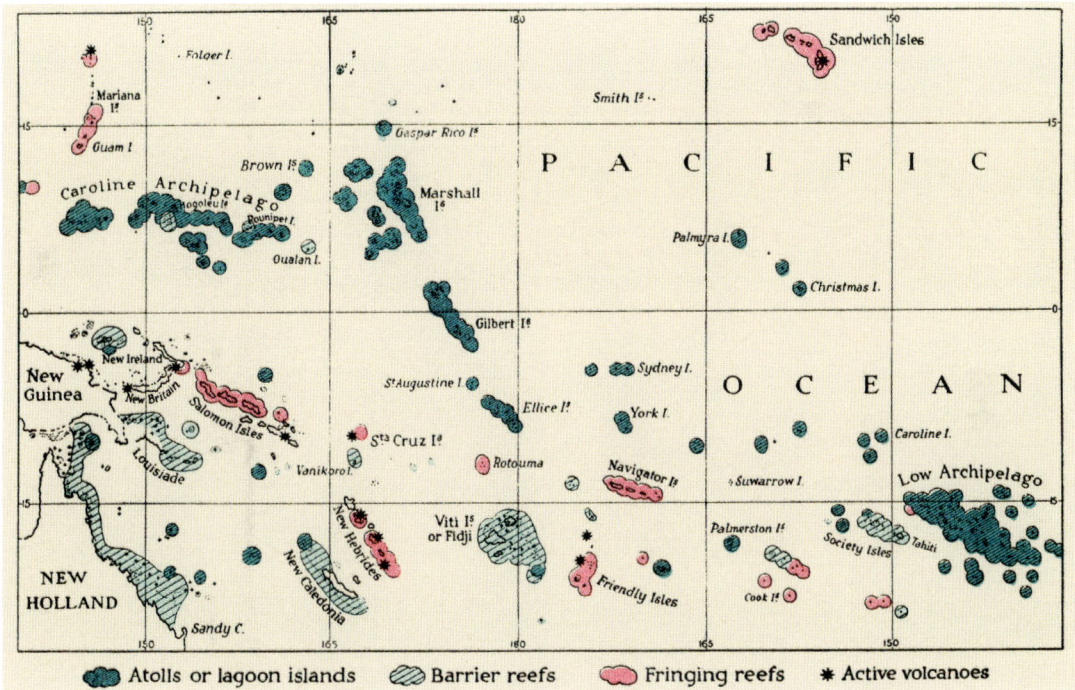

Charles Darwin (1809–1882), "Shewing the Distribution of the Different Kinds of Coral Reefs, Together With the Position of the Active Volcanos," from *The Structure and Distribution of Coral Reefs*, 1842.

Fascinated by the idea that the Earth's surface could rise and fall over long periods of time, Darwin meticulously mapped the world's coral reefs. It had been known for some years that coral is formed by tiny animals which can only grow in shallow water. In his 1842 *The Structure and Distribution of Coral Reefs*, Darwin proposed that ring-shaped coral atolls form around oceanic volcanoes after their summits are thrust clear of the surrounding waters. Uncharacteristically disagreeing with Charles Lyell, Darwin proposed that the volcanoes then collapse and submerge, but that this happens sufficiently slowly that the coral organisms can build up their reef at the same rate the volcanoes are sinking, and thereby can remain close to the surface. Thus, as the volcano sinks away, a delicate ring of coral is left isolated in mid-ocean.

The later discovery of igneous rocks beneath some atolls lent credence to Darwin's theory, but it is now thought that atolls have a different origin—although it still relates to the coral's ability to grow as fast as the sea level rises around it. Over the last half a million years, several cycles of global glaciation have reduced the world's sea levels by as much as 120 m (400 ft)—exposing new shallow regions of seafloor which coral can colonize. Then, when the distant glaciers melt and the seas "re-fill," the corals keep pace, and build their ring-shaped edifices upward as the water rises around them.

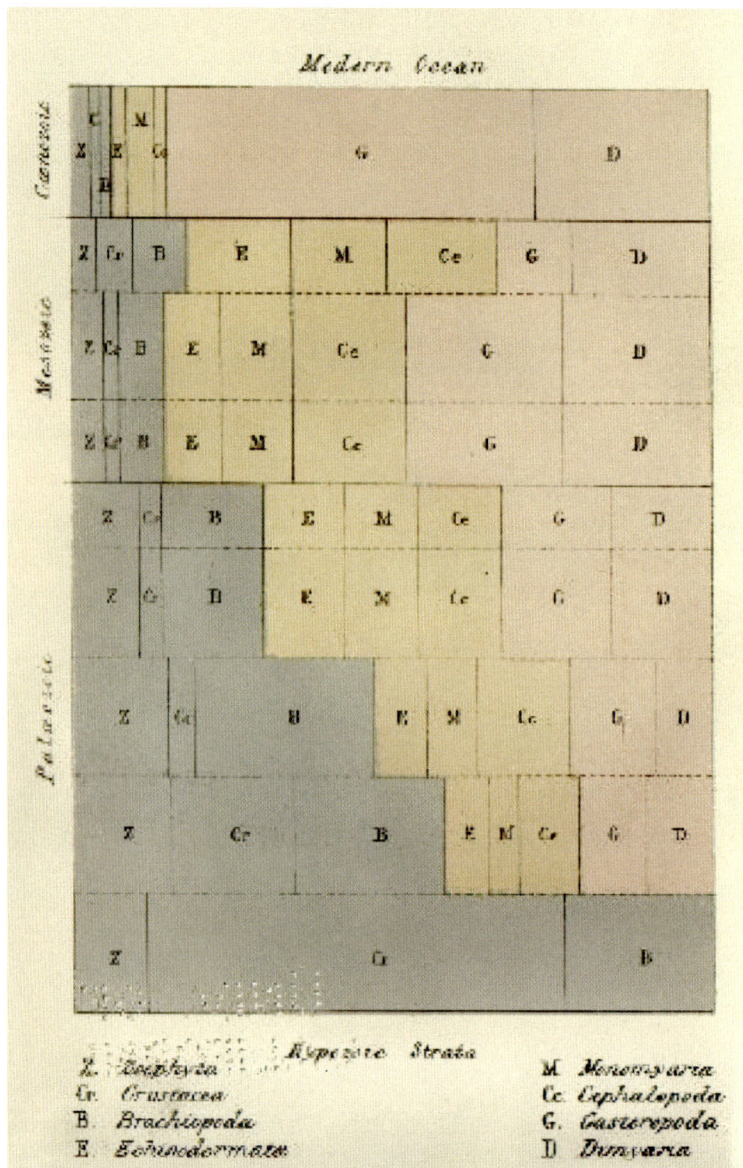

John Phillips (1800–1874), successive systems of marine invertebral life, from *Life on Earth and its Origin and Succession*, 1850.

Although the "tree" is the most common visual representation of the evolution of life, evolutionary trees are often rather divorced from the geological context in which fossils are found. It is difficult in a single image to give a sense of both the branching pattern of evolution and the layered arrangement of rock strata. In these mid-century "compromise" diagrams, the age of rocks runs from oldest at the bottom to youngest at the top, while the horizontal dimension indicates animal diversity and abundance.

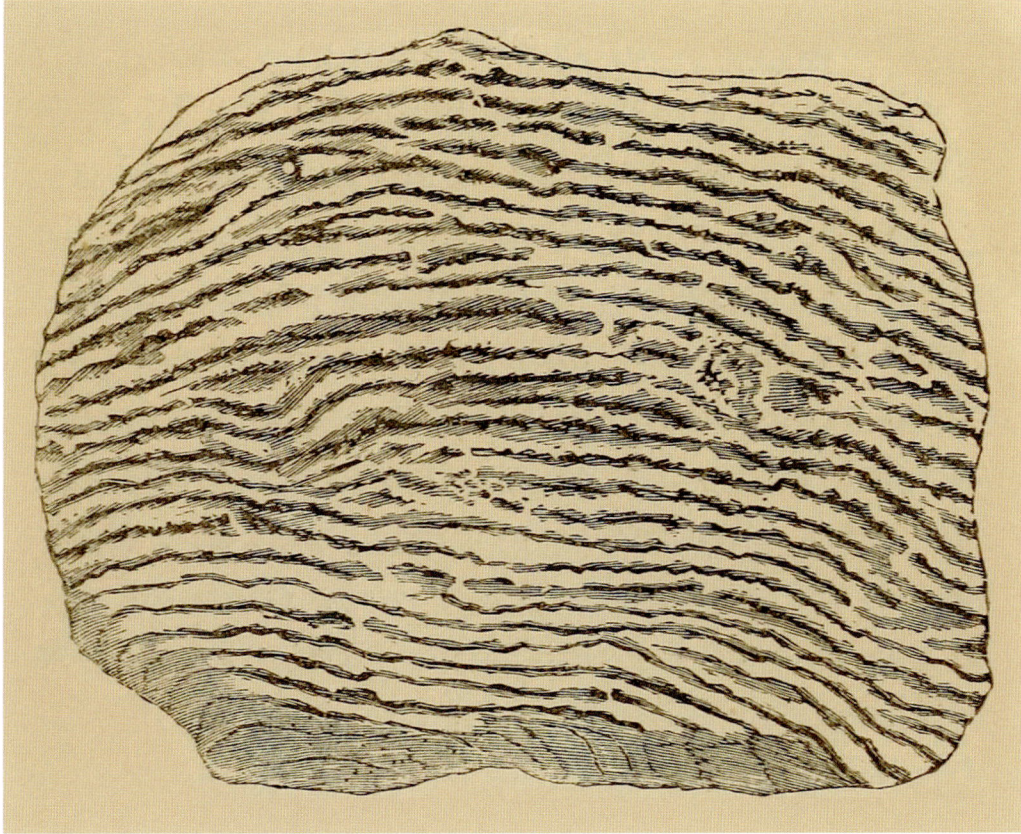

Alexander Logan (1841–1894), "Supposed Fossil from Laurentian Limestone, Grand Calumet," from *Geology of Canada*, 1863.

Sometimes fossils are not really fossils at all. Identified in 1850 and named, with some national pride, *Eozoön canadense* ("Canadian dawn animal") was found in rocks now known to be approximately 1000 million years old. The stripiness of *Eozoön* was claimed to result from deposition of silt in the chambers of its "shell," but soon paleontologists claimed the pattern was entirely mineral in origin, and drew parallels with the patterns seen in some types of marble. After decades of acrimonious argument, the paleontological consensus shifted to the conclusion that *Eozoön* is, instead, what is called a "pseudofossil"—it looks like a fossil, but isn't one.

LEFT *Archaeopteryx lithographica* fossil ("Berlin specimen"), Solnhofen, Southern Germany, 1874.

Lagerstätten, or sites which yield exceptionally preserved fossils, are rare things indeed. The limestone quarries of Solnhofen in Germany are particularly known as the source of all fourteen near-complete specimens of the most famous fossil of all, *Archaeopteryx lithographica*, "ancient wing drawn on stone." Some 150 million years ago, the sediments at Solnhofen were deposited in extremely fine layers which may now be split apart like the pages of a book, occasionally revealing the remains of ancient creatures etched onto the strata.

Archaeopteryx was first discovered in the form of a fossil of a single feather—an asymmetrical feather like those typical of the wings of modern bird species which can fly. Soon several whole-body fossils were also found, possessing a fascinating mix of reptilian and avian characteristics—the creature was feathered, proportioned like a bird except for its long tail, yet had clawed wings and teeth in its jaws. The collar bones and pelvis were halfway toward the distinctive arrangement of modern birds. *Archaeopteryx* neatly filled an important void in the history of birds: it is almost too perfect a "missing-link" fossil, and this led to claims it was faked, or artificially augmented.

ABOVE Great Rift Valley, Tanzania, 2015.

The paleontological study of humans started in 1891 with the discovery of the 0.7–1 million-year-old *Homo erectus* "Java Man." Dutch anthropologist Eugene Dubois (1858–1940) was convinced there must be fossil evidence of human "missing links," and this drove him to seek those links in the forests of the Dutch East Indies (now Indonesia).

However, since that time much of the focus of paleoanthropology has been on the African part of the 6,000-km (4000-mi.) Great Rift Valley which runs from Jordan to Mozambique. Rift valleys form when regions of continental crust are pulled apart by horizontal forces, in this case splitting the African tectonic plate in two. First described in 1893, the African Rift Valley is actually a tangle of faults around a complex central rift. This central depression bisects mountain ranges in the north, and in the south holds some of the world's most capacious lakes. The Rift Valley tells the story of six million years of human evolution, and was the location of the discovery of 3.3-million-year-old "Lucy" the *Australopithecus afarensis*, the earliest stone tools from 3.2 million years ago, and early *Homo sapiens* from around 230,000 years ago.

Some have suggested that abrupt transitions in human evolution were triggered when the volcanos which burst through the valleys' thinning crust underwent one of their periodic surges of eruption. Also, many think the Rift Valley was the origin of multiple waves of human migration which repeatedly spread to far-flung regions (including Java). However, others have suggested that human evolution took place all across the African continent, and that this giant sunken Lagerstätte is simply the place where human fossils were best preserved.

Taphonomy—The Law of the Grave

IVAN EFREMOV (1908–1972)

The raw material of paleontology is fossils, but although many thousands of fossils have been characterized, those are a vanishingly small and probably unrepresentative fraction of all the organisms that have ever lived. Very few organisms become fossilized after they die, and most fossils are later damaged or obliterated. And, only a minuscule minority of those which survive are discovered.

Early in the nineteenth century scientists were already worried about "gaps" in the fossil record. Sometimes entire communities of organisms seem to vanish suddenly at a particular stratum, to be replaced by new organisms or no organisms at all. Geologists such as William Smith (pages 216–217) questioned whether these "breaks" represented extinction events, or times when the processes of sedimentation simply stopped for some reason. The fossil record is fragmentary and capricious, and even if fossils from many diverse locations were to be gathered, it is questionable how complete the fossil story can ever be.

Even more concerning than the incompleteness of the fossil record is the possibility of *bias*. For example, mollusks and vertebrates are overrepresented in the geological and paleontological literature because shells and bones fossilize readily and robustly. However, there are many stages of the fossilization process at which more subtle biases can appear. For example, by the start of the twentieth century it was clear that running water "sorts" animal remains in surprisingly specific ways—certain bones may be selectively deposited at particular bends in a stream, and in some cases mirror-

Paleontologist and science-fiction writer Ivan Efremov.

Efremov was a central figure in geologists' and paleontologists' efforts to make the study of fossils more systematic. He argued that the fossilization, exposure, and discovery of organic remains is an erratic and biased process.

image bones from right and left sides of a body may be systematically separated.

In 1940, the Russian paleontologist and science-fiction writer Ivan Efremov coined a new term in *Taphonomy: A New Branch of Paleontology*. Literally meaning "the law of the grave," Efremov argued that a new discipline was needed to address the fact that fossilization is, in his words, "desultory," "incomplete," and "casual." He concisely defined taphonomy as the study of how the remains of organisms transition "from the biosphere to the lithosphere," (i.e. from the zone of living matter on the planet's surface, to its crust.)

"Necrolysis" is the first stage of fossilization, when some parts of a dead body decay or are otherwise destroyed—for example, human fossils often lack fingers and toes because they might be nibbled by leopards. Sometimes bodies may be moved before they fossilize, perhaps being washed downstream, or sinking to the seabed. For biological material to fossilize it must be sheltered from scavenging, decay, and abrasion—and this often involves being buried under sediments deposited from a body of water.

Water also plays an important role in the next stage, when the biological remains are converted into a fossil. The most common mechanism by which this happens is "petrification" in which water percolates through and slowly deposits minerals which replace the organic material. Really this stage is the central part of Efremov's "biosphere-to-lithosphere" transmutation.

In some ways, taphonomy is the part of paleontology which is most geological and least biological. Biologists instinctively want to know the story of life, but geologists want to understand how that story was written.

"Biologists instinctively want to know the story of life, but geologists want to understand how that story was written."

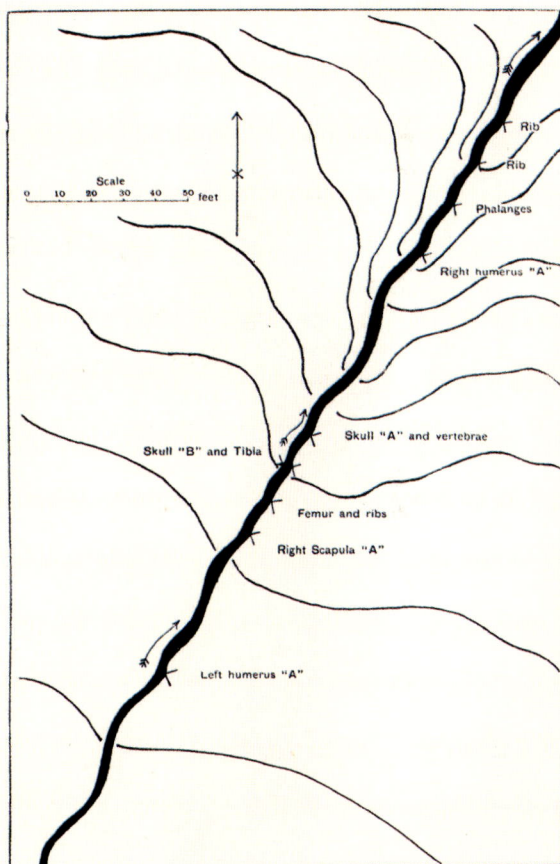

Wynfrid Duckworth (1870–1956), diagram showing the dispersive power of running water on skeletons, from *Studies from the Anthropological Laboratory: The Anatomy School*, 1904.

This map from an earlier study shows that Efremov was not the first to worry about the capricious nature of fossilization. Here a stream flows from top-right to bottom-left, and different bones from the same animal settle selectively at particular points along the flow of water.

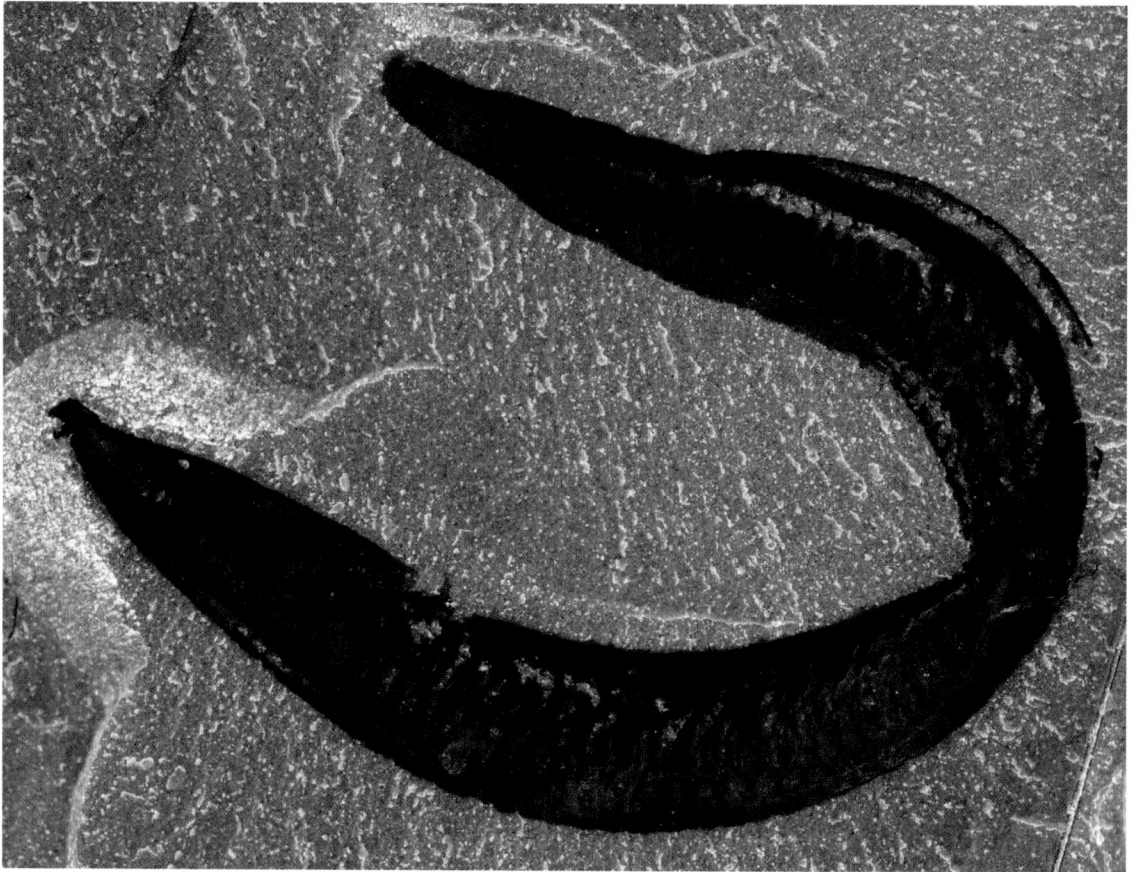

The Burgess Shale fossil *Pikaia gracilens*, discovered in 1909.

On August 30, 1909, Charles Walcott (1850–1927), Secretary of the Smithsonian Institution, stumbled upon perhaps the most important fossil Lagerstätte of all.

A particular 2-m- (6½-ft-)thick shale bed at the Burgess Pass near Mount Stephen in British Columbia is 509 million years old, twice as old as the oldest dinosaur, yet it contains a spectacular array of varied, delicate fossils of entirely soft-bodied creatures. Over the two decades after its discovery the "Burgess Shale" yielded hundreds of fossils, many quite unlike anything found elsewhere in the fossil record. Indeed the fossils' strangeness caused them to be misidentified for decades, as paleontologists forced them to fit into the major groups of animals we see today. Instead we now think these creatures represent a time when animals were "experimenting" with a variety of body plans—some of which later transformed into modern forms, but others were never seen again. This image is of *Pikaia*, a finely-segmented wriggly creature which may be related to the ancestor of vertebrates.

The term "Cambrian explosion" is sometimes used to refer to the bewildering variety of animals present in the shale, but some geologists think this is misleading. The Burgess Shale is just a snapshot, and we have so few fossils from the tens of millions of years before and after, that we simply do not know how sudden the diversification of early multicellular life was, nor whether any of the creatures Walcott discovered actually left any descendants.

Stephanie Pierce et al., illustration from "Three-dimensional limb joint mobility in the early tetrapod *Ichthyostega*," *Nature*, vol. 486, 2012.

One of the most dramatic transitions in the history of life was the colonization of land by vertebrates. Invertebrates might have got there first, but as vertebrates are larger in general than invertebrates, their size meant they faced particular physical challenges as they adapted to life on land. The modifications they needed were profound—fins became jointed limbs, gills were superseded by lungs, eyes and ears changed to operate in air rather than water, and the spine grew stronger to absorb the twisting movements inherent in walking on dry land.

This is a reconstruction of *Ichthyostega* from 360 million years ago, discovered in 1932 and the first fossil which seemed to be near the evolutionary lineage which brought vertebrates onto land. Greenland has been a particularly productive source of these early four-footed "tetrapods," including the evocatively named *Acanthostega*, *Tiktaalik*, and many of their more fishy relatives. The evolution of early tetrapods was a slow process—it seems to have taken many tens of millions of years. Indeed, the oldest fossil evidence of land vertebrates is a set of purposeful footprints left on a beach in what is now Poland, some 395 million years ago.

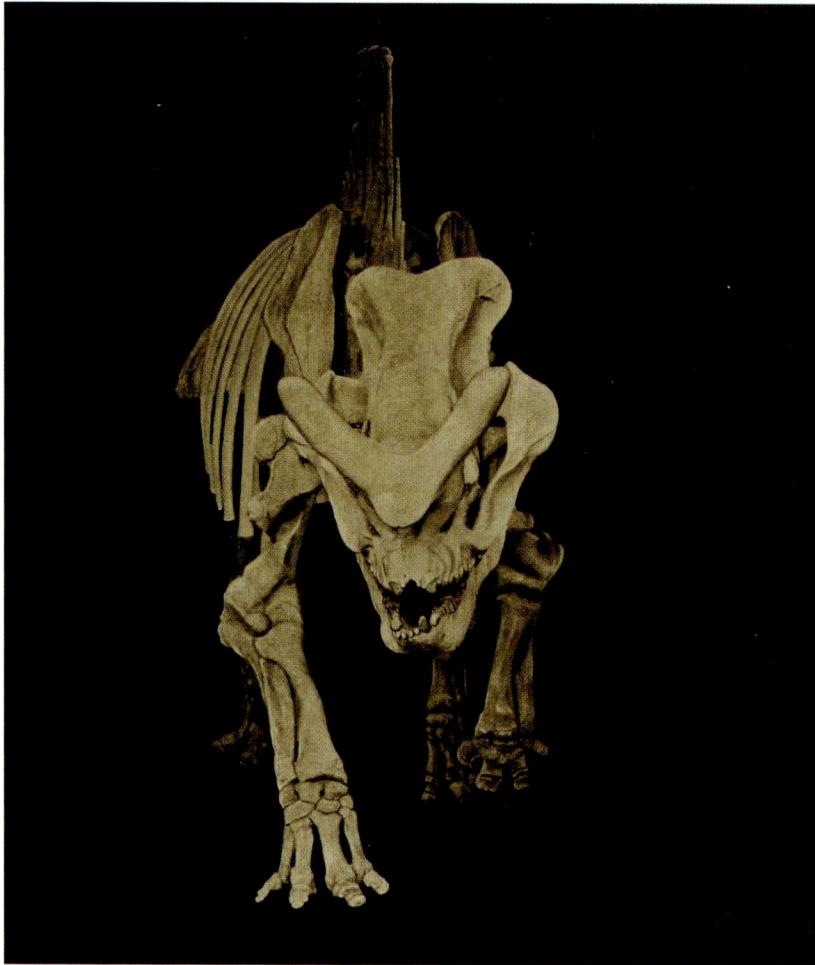

Henry Fairfield Osborn (1857–1935), "Mounted skeleton of *Brontotherium Hatcheri*," from *The Titanotheres of Ancient Wyoming, Dakota and Nebraska*, 1929.

The late eighteenth and early nineteenth centuries was the "classic" period of American paleontology, when many of the most famous dinosaurs and fossil mammals were discovered. There was a paleontological rush to the West, as East Coast institutions and philanthropists funded expeditions to dig the great new fossils from the ground and ship them back east, literally by the railroad carload.

The driving force behind many of these expeditions was the patriarch of New York's American Museum of Natural History, Henry Fairfield Osborn. Born into a wealthy family, Osborn controlled paleontology from above, funding and directing rather than participating in the actual extraction of fossils from the dirt. He named and published the resulting discoveries and left a voluminous scholarly legacy, especially on mammalian evolution and dinosaurs—he coined the names *Velociraptor* and *Tyrannosaurus rex*, for example. His work on the ancient mammals of North America was particularly important, and he delineated the evolution of the horse to a level of detail greater than any other species.

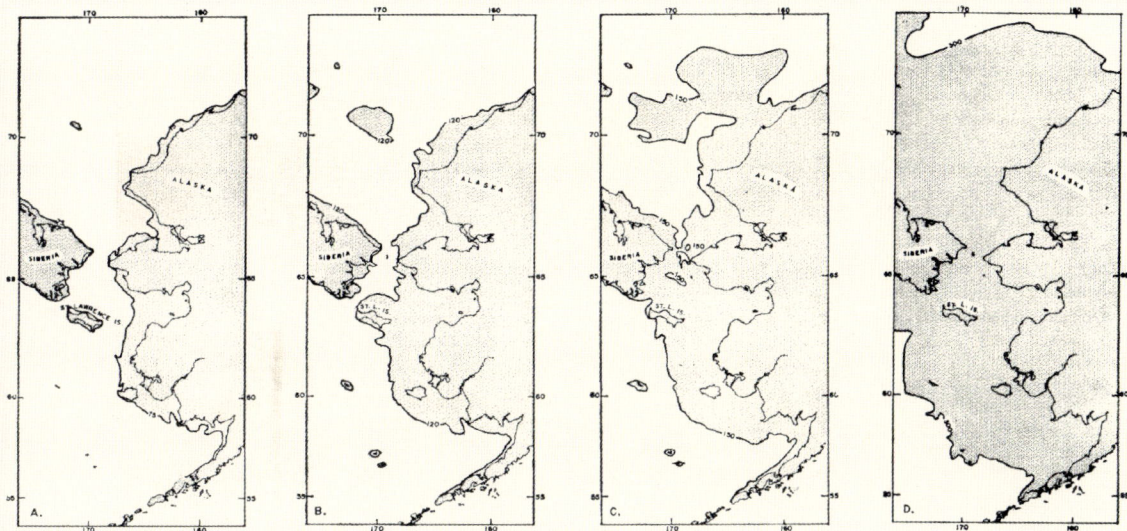

David M. Hopkins, "The probable outlines of the Alaskan and Siberian coasts when the sea level was 75 feet, 120 feet, 150 feet, and 300 feet below the present level," from *Cenozoic History of the Bering Land Bridge, Science* vol. 129, 1959.

We live at a time when the global distribution of the continents is particularly interesting. They are not all collected together into a single mass, nor are they so scattered that they are never in contact. Instead, they are tantalizingly arranged such that when sea levels are low, Africa, Eurasia, and the Americas are interconnected, but when sea levels are high, they are not. The most important element of this intermittent connectivity is the Bering Land Bridge which, when sea levels fall, connects Eastern Siberia to Alaska.

The Spanish missionary and naturalist Fray José de Acosta (1540–1600) was one of the first to suggest that Asia and North America were connected in the north, and he speculated that this was how the Americas had been peopled. We now know that the two continents are not currently joined but that humans have existed in the Americas for at least 16,500 years. It was the twentieth-century geologist David M. Hopkins who revolutionized our thinking about the Land Bridge. He showed clearly how a relatively small drop in sea levels would expose dry land between the continents, and also that the fauna and flora on either side of the Bering Strait exhibit clear signs of having been "in contact." And, complementing this, he argued that marine mollusk populations of the Arctic and Pacific Oceans were separated at times by a land barrier.

Life, but not as we know it

STAR TREK AND DOCTOR WHO

Works of television science have crystallized some intriguing interactions between geology and life. Notable among these is the discovery of the "Horta" in the "Devil in the Dark" episode of of *Star Trek*. In this tale our heroes meet blobby alien beings whose biology is based on silicon rather than carbon, and who burrow through solid rock as easily as a human would walk through air. Despite the biological and cultural barriers, Spock is able to mind-meld with a lady Horta and discovers that she finds that "our appearance is revolting, but she thought she could get used to it."

Indeed, scientists have long speculated about whether life can be based on silicon rather than carbon. Silicon sits immediately below carbon in the periodic table (page 39) and shares some of its properties. Silicon and carbon atoms usually form four bonds with other atoms, and this provides more scope for complexity than the one, two or three bonds which many other elements form. Because of this, silicon can combine with a wide variety of elements. It is also more plentiful in many celestial bodies' crusts than carbon.

Silicon has its downsides, however. It is not as good for making complex molecules as carbon, which may limit

silicon-based life's ability to extract energy from its environment. Silicon is also less reactive than carbon which would definitely slow things down. Finally, while the waste product of an oxygen breathing carbon based organism is carbon dioxide gas, for a Horta would this be scratchy silicon dioxide sand?

A different alternative history of life was posited in the 1970 series "Doctor Who and the Silurians," which has given its name to a hypothesis lying at the interface between geology and biology. In the TV story, Doctor Who encounters the Silurians, an ancient native species lying dormant under the Earth's surface, but now reawakened. The eponymous "Silurian hypothesis" is that humans were not the first technological civilization on Earth, but are in fact the second—preceded millions of years ago by a race which has now gone extinct. If this were the case, how would we know?

For a civilization to be detectable after hundreds of millions of years it must leave a persistent geological signature.

Ocean crust is usually destroyed within tens of millions of years (page 90) so any evidence must be on land. Fossilization is an extremely rare event (page 224) and we assume that most human bodies and technological artifacts will be obliterated quite quickly. Human activities alter rates of sedimentation and pollute the environment, but these effects may be indistinguishable from natural geological processes. Energy production and industrial processes have spewed nonnatural mixtures of isotopes (page 45) into the environment, and a few of these may be detectable for hundreds of millions (plutonium) or even billions of years (uranium). Finally, we currently have no evidence that someone burnt all the coal and oil before we found it.

LEFT **Spock approaches the Horta, 1967.**

Silicon-based life entered the public imagination largely due to the *Star Trek* episode "Devil in the Dark," in which miners on a far-flung planet accidently provoke a very protective silicon-based mother.

RIGHT **A Silurian, 2013.**

First appearing in *Doctor Who* in 1970, the Silurians have enjoyed a long on-screen career. Their original resurrection from the crust of the British Isles popularized the "Silurian hypothesis" in which geologists speculate about what evidence we might find had another technological civilization preceded us on Earth.

Banded iron formations: layers of iron-rich and silica-rich sediments that settled on an ancient sea floor 2,700–2,400 million billion years ago.

If an alien were to examine the Earth from afar, the planet's most striking feature would be the amount of oxygen in its atmosphere—far more than could be explained by geological processes. Almost all this oxygen comes from photosynthesis by living organisms, and this implies that long ago the Earth was largely oxygen-free.

The study of the Earth's early atmosphere started in earnest in the late 1960s. The type of photosynthesis carried out by most modern plants was first evolved by "cyanobacteria," single-celled organisms which use the energy from sunlight to convert carbon dioxide and water to sugars and oxygen. Cyanobacteria started to raise atmospheric oxygen levels from 2,400 to 2,100 million years ago in what has been called the "Great Oxygenation Event."

Much early oxygen became dissolved in sea water, but eventually it reached the atmosphere and started to react with minerals on the exposed surface of the continents. Evidence of the oxidation of minerals under water survives as "banded iron formations" like that in the image. Because oxygen was initially absorbed by rocks in this way, it may have constituted only 2 percent of the atmosphere at that time. However, that small amount was enough to kill most life on the planet and remove so much methane (CH_4) from the atmosphere that it cooled down, plunging the Earth into prolonged severe glaciations.

Fossil of the late-Carboniferous dragonfly *Meganeura monyi*, 2019.

After further fluctuations, oxygen concentrations rose in earnest from 800 to 400 million years ago when the gas came to constitute at least as much as the 20 percent of the atmosphere it does today. So although life produced the oxygen in the first place, and almost killed itself in the process, the presence of oxygen meant that organisms could now use it to "burn" their metabolic fuel to produce far more energy. This energy boost meant cells could become large, complex, and clump together to become multicellular organisms. Suddenly life was no longer just microscopic "bugs."

Later, the colonization of the land by multicellular plants led to yet another surge in oxygen production, and it is possible that from 350 to 300 million years ago, oxygen may have made up as much as 35 percent of the air. From the early twentieth century, scientists argued that the huge insects present at that time, including the dragonfly *Meganeura* pictured here, which had a 70-cm (27½-in) wingspan, thrived because of these high oxygen levels. At today's oxygen concentrations, insects' breathing systems usually only work well when they are small—which is why modern insects are small.

ABOVE **Morning Glory Pool, Yellowstone National Park, Wyoming, 2015.**

LEFT **Hydrothermal vents in the Pacific Ocean, 2024.**

Two of the basic requirements for life are energy and chemical nutrients. We have no particular reason to assume the first life on Earth was powered by solar energy, and most scientists believe that atmospheric oxygen was not involved, either. Indeed, discoveries over the last half-century have shown that life can survive, and may even have started in exotic, extreme geological locations.

In 1976 ocean surveys detected unexpected temperature spikes in deep ocean waters—with regions of water as hot as 400°C (750°F) scattered across the icy depths. Using submersibles, rocky "chimneys" were discovered spewing hot, dark water into the deep sea. These "hydrothermal vents" have now been discovered around the world, where water is heated by subsurface magma, is ejected, and deposits its dissolved minerals to form tall, tubular structures. Most remarkably of all, the noxious mix of chemicals is exploited by bacteria living around the vents, and which themselves become food for communities of deep-sea animals.

Consequently, hydrothermal vents are ecosystems which function entirely without solar energy, and they now even serve as models of how life may have started on Earth. We think there are two main types of vent—those which are hotter than 100°C (212°F) and those which are cooler—and the cooler ones are powered not by heat from magma, but by geological chemical reactions called "serpentinization." This reaction occurs when certain magnesium/iron minerals react with water to produce hydrogen (H_2) and methane (CH_4). These reactions are similar—suspiciously similar some would say—to chemical reactions employed by some bacteria.

The more one looks, the more one finds "extremophile" organisms living in apparently harsh conditions. Geysers and geothermal pools are often surrounded by colored rings of mineral deposits and "thermophile" bacteria which thrive in hot water. Elsewhere, some fungi seem to seek out radioactive materials, and it has even been claimed that some species living in the reactor at Chernobyl are able to use ionizing radiation as an energy source.

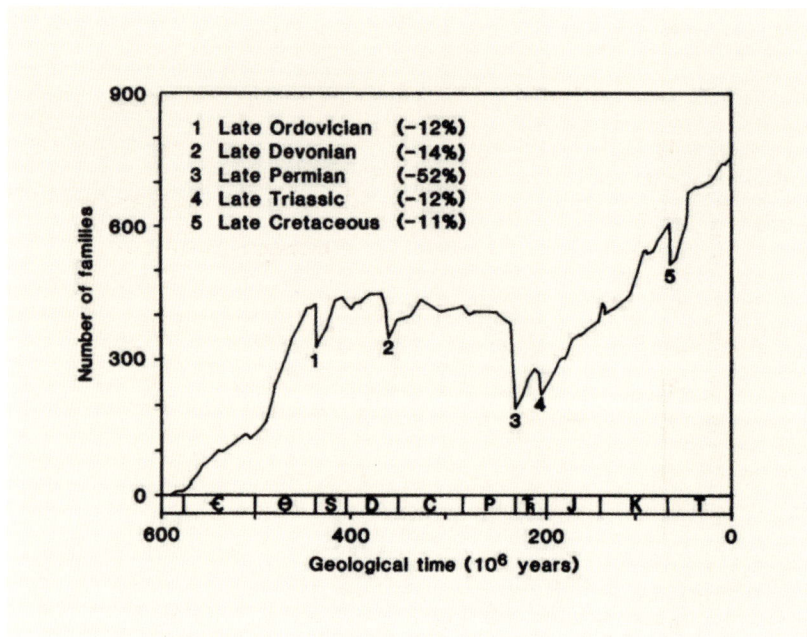

David Raup and John Sepkoski, graph showing standing diversity through time for families of marine vertebrates and invertebrates, from "Mass Extinctions in the Marine Fossil Record," *Science*, vol. 215, pp. 1501–3, 1982.

Since before the time of Cuvier (page 215) scientists have pondered mass extinction events. Are they a major or regular driver of evolution? Do they explain irregularities of the fossil record? What causes them, and are they all caused by the same thing?

This graph from 1982 shows marine fossil diversity over the last 600 million years (a "family" is a grouping into which organisms are classified). Four or five sudden major declines are evident with a clarity not seen before this paper was published. Extinction "5" was the late-Cretaceous event which famously caused the extinction of the non-avian dinosaurs, but extinction "3" was the most profound. This Permian–Triassic extinction event 252 million years ago is sometimes called the "Great Dying" because an estimated 80–96 percent of marine species were extinguished. It is thought the Great Dying was triggered by enormous eruptions of basalts which formed the "Siberian Traps," a "large igneous province" in northern Russia. Release of carbon dioxide and sulfur dioxide (SO_2) caused global warming, ocean acidification, and a decline in oxygen levels in seawater.

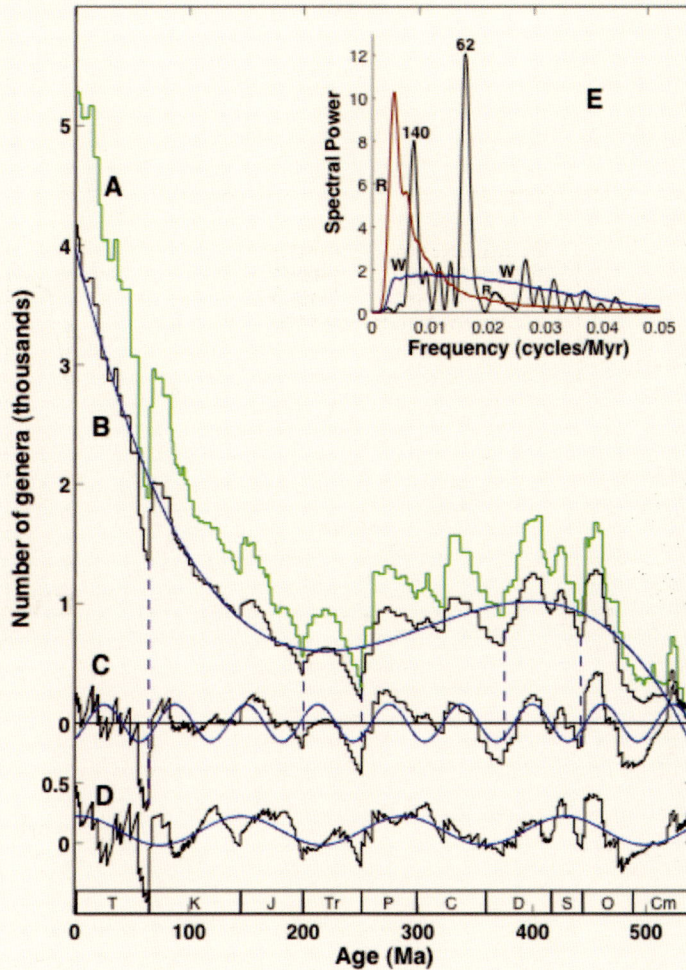

Robert Rohde and Richard Muller, graph showing genus diversity, from "Cycles in Fossil Diversity," *Nature*, vol. 184, pp. 208–10, 2005.

It is now agreed that the diversity of living species has fluctuated dramatically since the Cambrian period, and this is evidence that intermittent extinction events have taken place. This graph is from a study which sought to determine whether there is any pattern to extinction events.

The green line "a" is a graph of animal diversity (in this case "genera" rather than "families") over the last 540 million years. The horizontal scale is "millions of years ago," so the present is on the left and the past is on the right (the opposite orientation to the graph on the facing page).

Some poor-quality data were removed to produce "neatened up" trace "b." When very long-term trends (sweeping blue curve) were removed, the resulting plot "c" matches a 62-million-year undulating cycle (blue wavy line) quite well, and further manipulation suggests a 140-million-year cycle as well, "d." The inset graph shows further statistical analysis of these data, and once again picks out superimposed cycles of extinction at 62-million-year and 140-million-year intervals.

All this is striking evidence of a pattern behind extinction events, yet no convincing link could be made with potential causes of extinction. There were hints that vulcanism and glaciation might be involved, but no connection was found to sea level changes, extraterrestrial impacts, the Earth's orbit around the sun, or the sun's orbit around our galaxy.

ABOVE The Ik-Kil cenote on the Yucatan Peninsula, Mexico, 2017.

RIGHT Late Cretaceous period dinosaur footprints in the Cal Orck'o cliff near Sucre, Bolivia, 2016.

Not all major extinctions are caused by geological events. Sixty-six million years ago something caused the extinction of most of the planet's animal species, and famously wiped out all dinosaurs except their diminutive avian descendants.

In the 1980s, geological evidence emerged that the late-Cretaceous extinction may have had an extraterrestrial cause. The metallic element iridium is usually rare in the Earth's crust, but around the world sediments deposited around that time exhibit a transient hundred-fold increase in concentrations of the element—a phenomenon called the "iridium anomaly." This led the father-and-son scientists Luis and Walter Alvarez (1911–1988 and b. 1940) to propose that the late-Cretaceous event was caused by the impact of an iridium-rich asteroid.

Although the cause of the extinction may not have been geological, the key evidence for it certainly is. Multiple lines of evidence suggest that the Alvarezes' asteroid formed a 180-km (110-mi.) impact crater centered just off the Mexican coast near Chicxulub. Satellite radar has detected a large buried bowl-shaped structure, and there is local evidence of melted and "shocked" minerals as well as damage to seabeds and beaches by massive tsunamis. The underground crater now interrupts the flow of groundwater, so that its rim is picked out by a telltale arc of cenotes, or sinkholes, similar to the one pictured here.

"Ichnofossils" are relics of animals' activities, not their bodies— "ichno" means "track." It may seem unlikely that footprints or burrows would fossilize well, but a single animal leaves many, many traces during its lifetime, so there is no shortage of fossilizable vestiges. As well as informing paleontologists about how animals behaved, these traces can be as distinctive as body-fossils for geologists trying to establish the stratigraphy of a sedimentary formation. Indeed, many extinct species are known *only* by their traces, while fossils of their bodies have never been found.

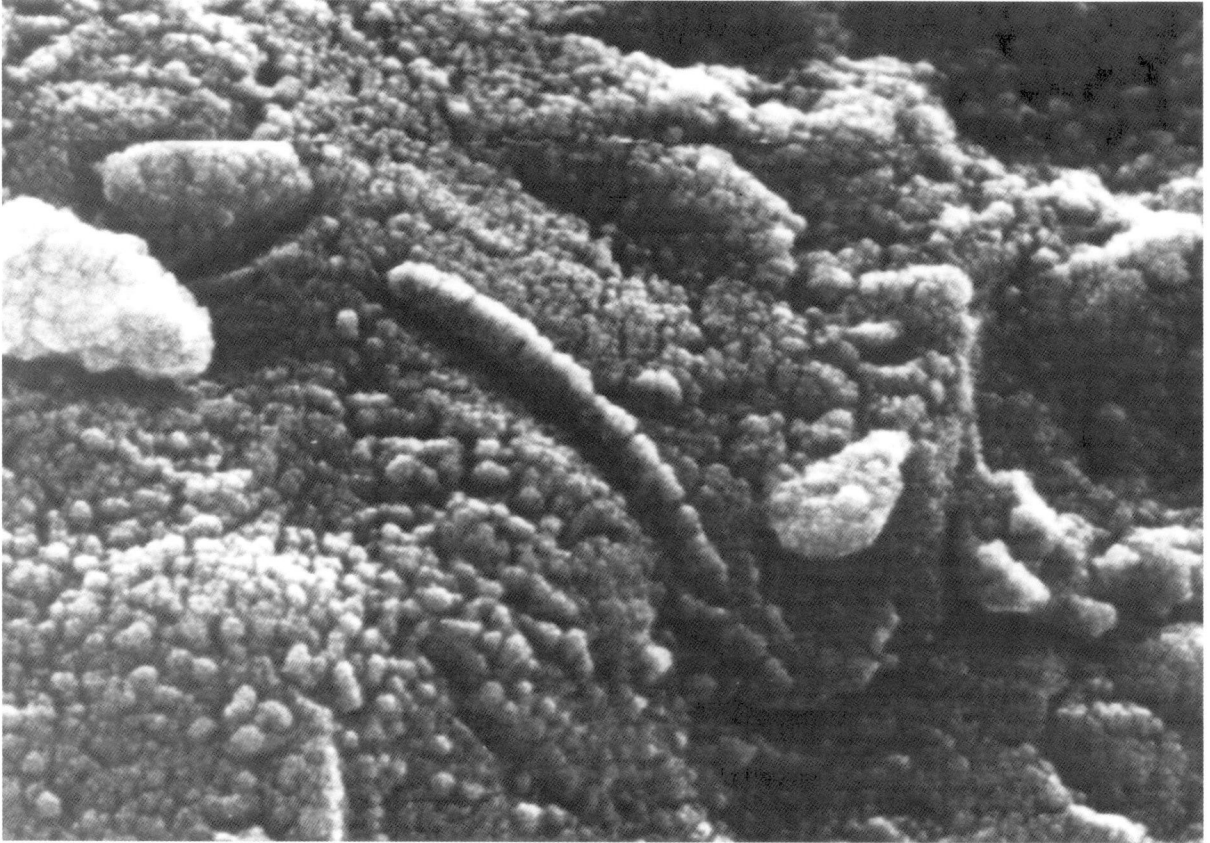

Electron micrograph of structures resembling organisms in the Mars-origin
Allan Hills 84001 meteorite, 1996.

Life on Mars? No potential extraterrestrial home for life has captured the
imagination as much as the Earth's small ruddy neighbor. Ever since Giovanni
Schiaparelli (1835–1910) misinterpreted Martian splotches of surface color as linear
channels, and his Italian *canali* was mistranslated into English as the very artificial
sounding "canals," many alien seekers have focused on the red planet.

However, more than once our hopes have been dashed, or at least dampened.
In 1976 the first spacecraft to successfully land on Mars, Viking 1 and 2, each
carried three experimental packages designed to detect life in the planet's soil.
One of these experiments yielded potentially positive results, but later probes have
suggested that chemical reactions of Martian minerals probably caused this result.

Perhaps the greatest disappointment came in 1996 when, with great fanfare, the
announcement was made that microbial fossils had been found in the "Allan Hills
84001" meteorite, a chunk of rock blasted from the surface of Mars and which
subsequently landed in Antarctica. Electron microscopy revealed minuscule
elongated entities within the meteorite, similar to rod-shaped *Bacillus* bacteria
on Earth, and it seemed that extraterrestrial life had at last been found. However,
it was soon agreed that these tiny "organisms" were in fact mineral in origin.

NASA Curiosity Rover multi-exposure "self-portrait" on the Martian surface, March 16 and 25, 2021.

One of the most confusing episodes in the search for life on Mars has been the contentious detection in 2012 of methane (CH_4) by the NASA Curiosity rover. On Earth methane is often produced by biological processes, so this finding drew a great deal of attention. Particularly noticeable is that the amount of methane fluctuates between Martian night (more methane) and day (little methane) and across the planet's seasons—at some times reaching concentrations forty times higher than at others.

Something about Martian methane does not make sense, however. Sensitive orbiting spacecraft have not detected any methane, and just to confuse matters some Earth-based telescopes have detected methane plumes covering large regions of the planet. Once ejected into the atmosphere, the gas should persist for a matter of centuries, not come and go according to the season or time of day. It has been suggested that Curiosity's weight is squeezing out otherwise resolutely subsurface methane when it trundles across the ground. And even if methane is present, geologists have argued that it could just as well be produced by mineral processes such as serpentinization (page 235).

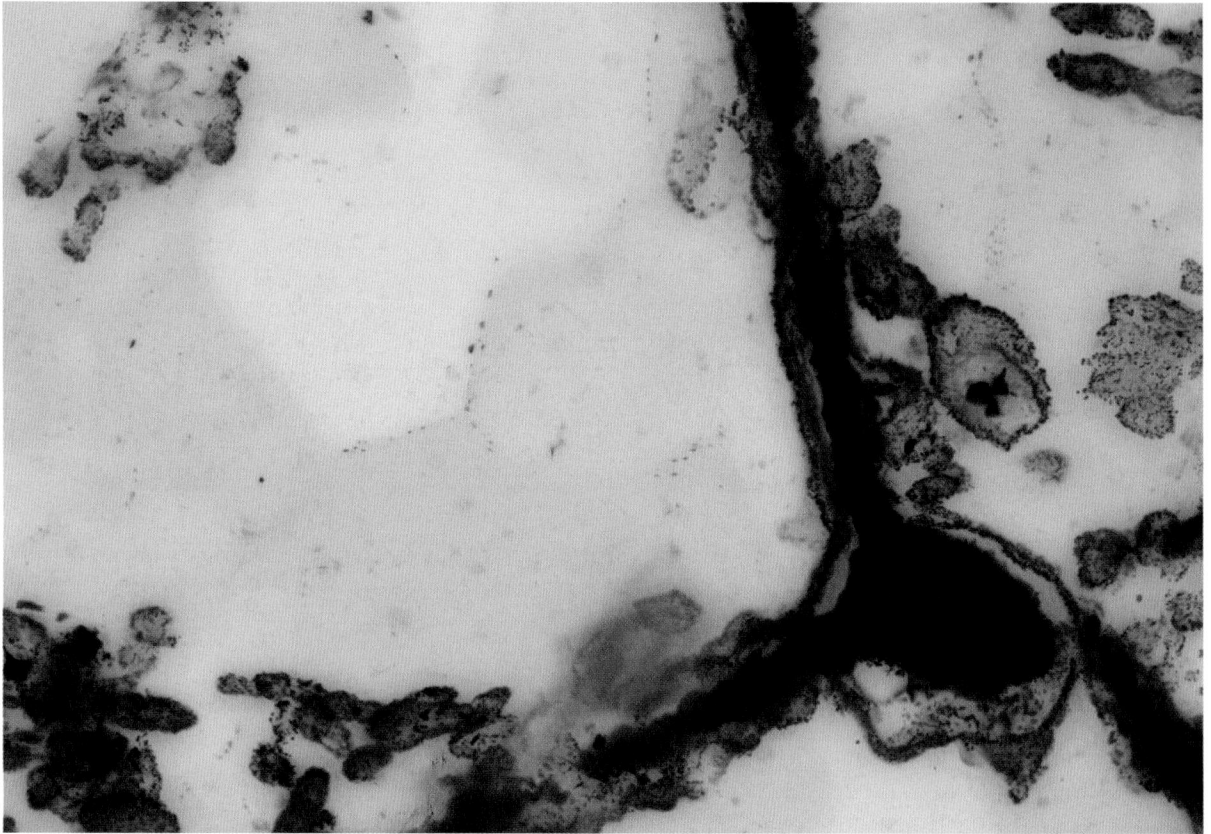

Matthew Dodd et al., transmitted light images of haematite filaments from the NSB and Løkken jaspers, from "Evidence for Early Life in Earth's Oldest Hydrothermal Vent Precipitates," *Nature*, vol. 543, pp. 60–64, 2017.

This image may be the earliest evidence of life on Earth. Few things in geology and paleontology are more contentious than claims to have found the earliest signs of life. For some time after its discovery in 1909 the 509-million-year-old Burgess Shale was something of a mark in the sand (page 226). From the 1940s the gradual acceptance of the validity of the Ediacaran biota extended the antiquity of multicellular life to around 630 million years (page 54). However, by then it was clear that slabs of fossilized cyanobacteria (page 232) called "stromatolites" were around 1,000 million years old, and possibly up to an astounding 3,500 million years. Presumably they were present by the time the levels of atmospheric oxygen began to rise.

This image is from a 2017 scientific paper in which the authors describe evidence of fossilized microorganisms in sediments from the Nuvvuagittuq greenstone belt in Quebec, deposited around undersea hydrothermal vents (page 235) as early as 4,280–3,770 million years ago. The fossils are spindles of the mineral haematite (an iron oxide, Fe_2O_3) and altered levels of carbon isotopes similar to vestiges of microbes living near later and present-day vents. If the Earth is 4,540 million years old, these findings would mean that life existed on our planet surprisingly soon after if formed. Indeed, it would have existed perilously close to the cataclysm of the Late Heavy Bombardment (page 50).

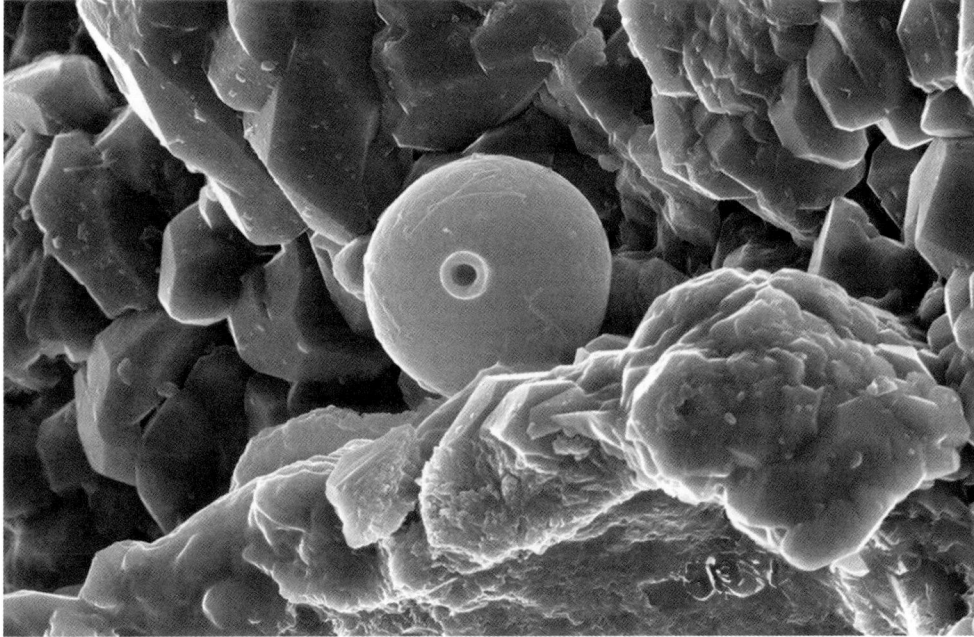

Andrea Hidalgo-Arias et al., scanning electron micrograph of a chrysophyte [golden alga] cyst in a volcanic rock sample, from "Adaptation of the Endolithic Biome in Antarctic Volcanic Rocks," *International Journal of Molecular Sciences*, vol. 24, 13824, 2023.

This image may be a premonition of the last life on Earth. In the long term, the future of life is not a hopeful one. In approximately 1,000 million years' time, three changes will conspire to extinguish all complex life from the surface of the Earth. By that time, a large fraction of the oceans may have been subducted (page 100) into the mantle. Second, increased levels of solar radiation will have increased the planet's average surface temperature to around 50°C (122°F). Worst of all, atmospheric concentrations of carbon dioxide will decline so much that photosynthesis will no longer be possible—no more oxygen will enter the atmosphere and all complex life will die.

The last living things will probably be microbes living deep below the surface and which do not use oxygen for their metabolism. Between 1,500 and 2,500 million years from now the very last organism to die will probably be "endolithic," living *within* rocks. Endolithic microbes are widespread across Earth today, although we are not sure how much of the world's biomass lives between the mineral grains of rocks in this way. They often employ "exotic" metabolic systems which do not use oxygen and can thrive in remarkably inhospitable places.

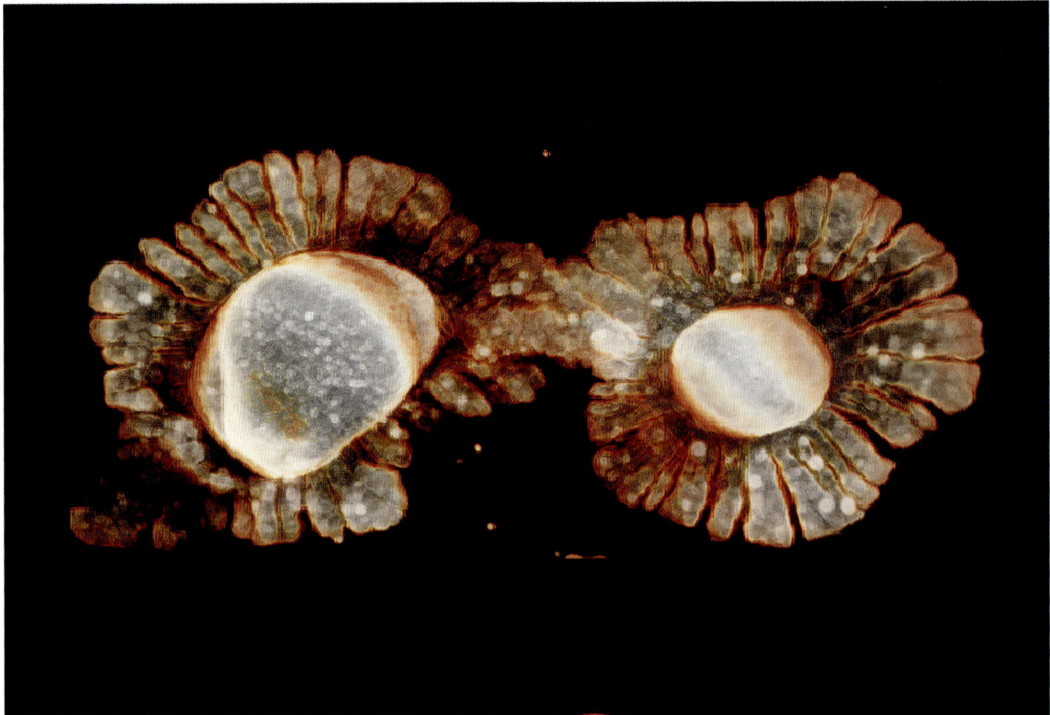

Abderrazak El Albani et al., lobate forms showing sheet-like structure and radial fabric, from "The 2.1 Ga Old Francevillian Biota: Biogenicity, Taphonomy and Biodiversity," *PLOS ONE*, vol. e99438, 2014.

The earliest fossil-like structures claimed to represent animals were found in the black shales of the Francevillian formation in Gabon in 2010. A variety of flowery, lobed, and elongated structures have been interpreted to be 2,100-million-year-old animals—just after the Great Oxygenation Event (page 232). If the "Francevillian biota" really were animals, then they would have needed oxygen. Animals, plants, and fungi are "eukaryotes"—made of cells more complex than those of bacteria, with nuclei to contain their genetic material, and usually needing a supply of oxygen. The identity of the Francevillian biota remains controversial, but some geologists think there is evidence that oceanic carbon, nitrogen, phosphorus, and iron were involved in processes which hint at the influence of life around this time.

Qing Tang et al., cellular structures of *Proterocladus antiquus* new species, from "A One-Billion-Year-Old Multicellular Chlorophyte," *Ecology & Evolution*, vol. 4, pp. 543–9, 2020.

A pivotal moment in the evolution of life on Earth came when a cyanobacterium (page 232) became engulfed by a eukaryotic host cell, creating the first plant. Like other eukaryotes, plant cells can join together to create large multicellular colonies and organisms. However, they have the additional advantage of using their cyanobacterial prisoners to derive energy from sunlight—those prisoners are now called "chloroplasts." This fossil is thought to be a 1,000-million-year-old tangle of algae. Plants are now the dominant life form on Earth—for example, perhaps 80 percent of the world's biological carbon is contained in tree trunks.

Where ocean meets rock

SWIMMING UNDER THE ICE AT JUPITER AND SATURN?

Over recent decades, the quest for life in the solar system has shifted in an unexpected direction. Many now believe that the moons of the giant planets Jupiter and Saturn, previously assumed to be frigid and sterile, offer our best chance of finding life in our cosmic back yard.

In 1979 the two NASA Voyager spacecraft sent back high-resolution images of Jupiter's fourth-largest moon, Europa, which showed a young, icy surface. These features, as well as Europa's exceptional smoothness, led to the idea that the moon possesses a deep water ocean below a thin icy shell. Remarkably, although Europa is slightly smaller than our own moon, its water layer may be so deep that it contains twice as much water as all of Earth's oceans.

Although the outer solar system receives much less solar radiation than Earth, this appears to be no barrier to these moons having active, heat-driven interiors. Many of the solar system's moons continually distort and recoil under the influence of gravitational forces from their parent planet. Internal friction and ocean currents generate heat which maintains the water in a liquid state, insulated by the overlying layer of ice.

And where there is water, scientists look for life. It may in fact be Europa's moderate size which makes it such a strong candidate, because it means its oceans are probably in direct contact with the rocks beneath. In contrast, in some larger moons such as the Jovian giant Ganymede, the pressure at the bottom of their larger oceans may be so high that a layer of dense "tetragonal ice" sits between the ocean and the rock underneath—forever separating apart the minerals and water which are two key raw materials for life.

In 1980 and 1981, the Voyagers visited the Saturn system and imaged the moon Enceladus. Only one seventh of Europa's diameter, it too turned out to have a smooth icy surface. Even more surprisingly, it continually spews water from multiple rifts in its southern regions, creating a ring of ejecta which orbits Saturn.

The discovery of the water plumes of Europa and Enceladus have changed our approach to searching for life beyond Earth. Instead of the difficulties and hazards of landing and drilling down to pristine oceans, we need only send spacecraft to fly past and "sniff" the ocean-water ejected from them—and indeed, on recent trips, NASA's Cassini spacecraft has detected cyanide, ethane, ethene, and methanol in Enceladus's plumes.

Evidence for these unexpected oceans continues to accumulate: observable plate-tectonic-like shifts in the icy skin of Europa, and electrical distortion of auroras by salty ocean currents. Yet it is the contact between these putative oceans and the geochemistry of the underlying rocks which could mean that one day, a small icy world far beyond Mars will be the site of the most important discovery in human history.

LEFT **True-color image of the Jovian moon Europa, NASA Galileo mission, 1995 and 1998 (composite image).**

Europa represents perhaps our greatest chance of finding life in our solar system. Compelling evidence suggests that its immensely smooth icy surface is a shell overlying a deep briny ocean, in contact with the satellite's rocky core.

RIGHT **Southern part of the Saturnian moon Enceladus, backlit by the sun, enhanced color image, NASA Cassini mission, 2006.**

This image of this small moon has been extensively processed to accentuate the plumes of liquid spewing from its south pole—in the "eight o'clock" direction. Ejection of water from these ice-and-water moons' internal oceans means that samples of those oceans will soon be sampled by spacecraft passing by, without the need to land and drill through the ice.

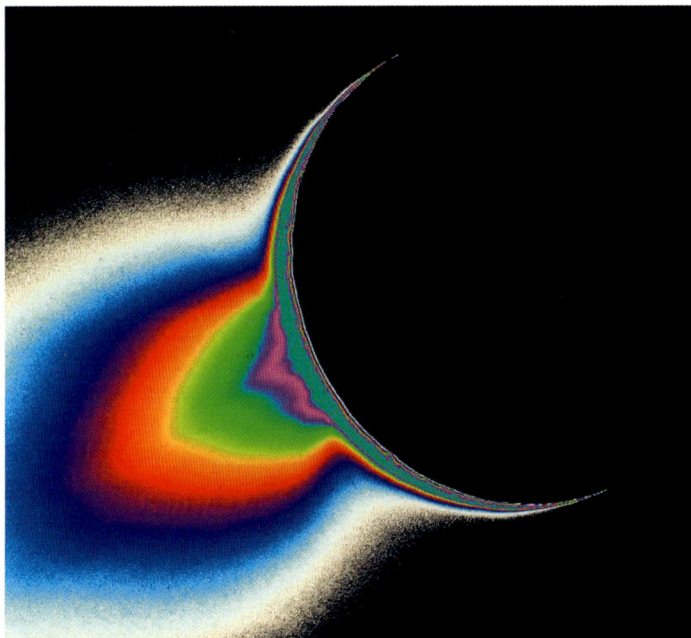

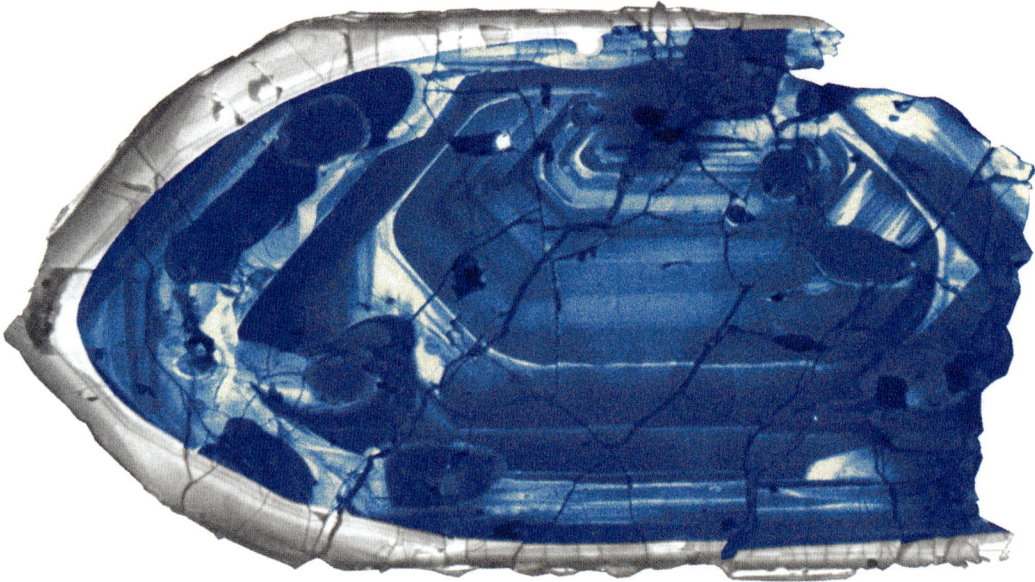

John Valley, false-color microscope image of a 4,400-million-year-old zircon from the Jack Hills, Australia, 2014.

We assume that liquid water was an essential prerequisite for the appearance of life on Earth. Because of this, geologists have tried to determine when liquid water first pooled on the planet's surface—after all, this defines the earliest time when life *could* have existed.

To do this, they returned to the earliest known minerals known to have formed on Earth, the Jack Hills zircons (page 52). If, as was argued in a 2024 study, the zircons formed when pre-existing crust minerals were exposed to liquid water brought to earth by icy meteorites, then water must have been available very early in Earth's history—at least 4,000 million years ago, and maybe even before that. Increasingly, it seems that our planet did not remain lifeless for very long.

Painting of a wild pig, from Leang Tedongnge in Sulawesi, Indonesia, made at least 45,500 years ago.

At the other end of the paleontological timescale, it is pleasing that the first glimmers of human visual art were based on very geological materials, and that their antiquity has been demonstrated by very geological techniques. For untold millennia caves have offered protection for humans and the artworks they left behind them, and the limestone karst (page 133) caves of the Indonesian island of Sulawesi contain some of the highest concentrations of prehistoric cave art in the world. In 2021 a research group published evidence that a depiction of a wild Sulawesi warty pig in one of these caves is the oldest known figurative art in the world. Produced using the mineral pigment ochre, the age of the painting was determined by reference to radioactive decay of uranium in the minerals that have been deposited on it since it was painted. The results suggest this ochre warty pig is 45,500 years old, almost exactly one one-hundred-thousandth of the age of the Earth.

Index

Page numbers in *italic* type refer
to illustrations.

Aberdeen Bestiary *21*
abiogenesis 203, 208–9, *209*
accretion 85
acid rain 133
Acosta, Fray José de 229
Adair, Paul 'Red' 192
Agassiz, Louis 111, 130, 131, *131*
Agricola, Georgius 116, *116*, 161,
    172, *172*
agriculture 159–60
alpha particles 177
Alps 82, 129, *129*, 131, 137
alumina 97
aluminium 39, 90, 143
Alvarez, Luis and Walter 238
Amargosa range *112*
amino acids 208
Ampferer, Otto 82, *82*
Anderson, Ernest 136, *136*
Andes 81
anisotropic 117
Anning, Mary 205
Antelope Canyon *12*
anticlines 121, 125, 177, 180
Appalachian Mountains 135, *135*
aquifers 178
Archaean 55
Archaeopteryx 222, 223
arches *11*, 133
Arduino, Giovanni
asteroids 19, 50, 55, 199, *199*, 238
asthenosphere 90, 104
Atacama Desert, Chile 189, *189*
atmosphere 55, 154, 200, 232–3, 243
atolls 34, 35, 219, *219*
atomic number 38
Aubert, Julien *105*

Ball's Pyramid 124, *124*
Bam earthquake 195, *195*
Barnes, H.T. 61
basalt 143, *146*, 200, *200*, 236
batholiths 124, *124*
bathymetry 89
bedrock 140, 150, 179
Beijing–Hangzhou Grand Canal 168,
    169
Bering Land Bridge 229
black smokers 164
Boltwood, Bertram 18–19, 45
Bouyahiaoui, B, et al *100*
Braer oil tanker 192, *193*

Breislak, Scipione *72*, 73
Bretz, J Harlen 140–1
Brocchi, Giovanni Battista 213, *213*
Brongniart, Alexandre *214*, 215
bronze age 166–7, *167*
Buffon, Georges-Louis Leclerc, Comte
    de 58
Burgess Shale 226, *226*, 242
Buridan, Jean 108, *109*
Büyük Menderes River 115, *115*

Cairngorm Mountains 120
calcite 170
calderas 32, *63*, 75
Callisto 50
Cambrian 43, 45
Cambrian explosion 228
canals 168, *169*, 173, 216
carbon 174–6, 180, 230–1, 244, 245
    sequestration 200, *200*
carbon dioxide 97, 119, 133, 161, 174,
    175, 180, 200, 231, 232, 236, 243
Carboniferous 43, 162, 175, *175*
cassiterite 165
catastrophism 17–18, 19
Cave of Crystals *152*, 153
caves 133, 249
cenotes 238, *238*
Cenozoic 27
ceramics 159–60
Chancourtois, Alexandre-Émile
    Béguyer de *128*
Chapeau de gendarme 125, *125*
Charpentier, Jean de 130, *130*, 131
Chaucer, Geoffrey 22
chemical compounds 161
Chernobyl 235
Cheyenne Mountain Complex 196, *196*
chloroplasts 245
chrysophyte *243*
cirques 111
clay 159–60, 161
Clerk of Eldin, John *28*
climate 52, 126
    climatology 41
    cyclicity 41
    global warming 163, 178, 200, 236
    ice ages 47, *47*, 104, 111, 131, 159
    Milankovitch cycles 47, *47*
    snowball Earth 50, *50*
coal 59, *159*, 162, 173–5, *174*, 183
Coalbrookdale 173, *173*
Columbia Plateau 140–1, *140*, *141*
comets 50
concrete 161
conflict minerals 188
conglomerates *44*, *45*, 153, *153*

continental crust 90, 92, 100–1, 104, 223
continental drift 47, 57, 62–5, 80–3, *80*,
    81, *83*, 86–9, *86*, *87*, 94, 229
    expanding Earth theory 86, 93
    mid-ocean ridges 63, 64, 81, *85*,
        88–90, 164
    orogeny (mountain-building) 81–2,
        82, *83*, 90, *91*
    plate tectonics 63–5, 82, 90–3, *91*
    supercontinent cycle 92, *92*
    thermal contraction theory 93, *93*,
        128
continental shelves 104, 175
continents 55, 64–5
    distribution 68, 80
convection 40
copper 164, *164*, 166, 167, 184, 195
Cordier, Louis 61
core 40, 58, 61–2, *61*, 68, *68*, 78–9, *78*, *79*
    core-mantle boundary topography
        102, 103, *103*
    density 62
    energy release 104
    inner 104
    magnetic field 104, *105*
    outer 104
    size 62, 104
    temperature 61, 104
Cornwall 165, *165*, 189
corries 111
coseismic interferogram *195*
cosmic radiation 209
coulees 140–1, *140*
Crameri, Fabio *85*
craters 48, *49*, 50, *51*, 138, *149*, 198, *198*,
    238
Creation narrative 14–15, 20, *21*
Cretaceous 25, 43
Croll, James 47
    Climate and Time 41, *41*
crust 61, *61*, 64–5, 68, *68*, 82, *126*
    continental 90, 92, 100–1, 104, 223
    distortion 112
    isostasy 81, 101, 104
    oceanic 64–5, 88–9, *88*, 90, 104, 231
    plate tectonics *See* continental drift;
        plate tectonics
    thickness 101, 103, 104
    thrust faulting 113, 129, 137
    tidal effects 94
    uplift 29, 34–5, 37, 112–13, *113*, 121,
        127, 135, 150, 164
crystals 109, 110, 117, 119, *152*, 153
    anisotropic 117
    atomic structure 139
    chemical formulas 119
    color 122

crystal lattice 119, 122, 176
diamond 176, *176*
shapes 119
Steno's Law 117
X-ray crystallography 139, *139*
Cuvier, Georges 18, *214*, 215, *215*
Cvijić, Jovan 133
cwms 111, *127*
cyanide 247
cyanobacteria 232, 242, 245
Cyprus 164, *164*, 167

Dalton, John 38, *38*
Daly, Reginald Aldworth 48
Darwin, Charles 33, 34, *34*, 41, 127, 206–7, 218–19
 The Structure and Distribution of Coral Reefs 219, *219*
Dauphin, Lauren 98
Davis, William Morris 112, 135, *135*
deltas 112, 114, *114*, 135, 175
deposition 22, 23–4, *24*, 113, *113*, 114, 221
 fossils *See* fossils; sedimentation
 measuring 14
 moraines 131, *131*, 150
 ore seams 172
Descartes, René 58, *58*, 68, *68*, 78
Devonian 43
diamond 176, *176*
Dickinsonia 54, 55
differentiation 68, 85, 124
diffraction 139
dinosaurs *202*, 203, 205, *206*, *228*, 236, 238, *239*
discontinuities 17, 29, 78
displacement theory 80–1
divining *172*
Dokuchaev, Vasily 179, *179*
dolomite 170
Dolomite Mountains 35
domes 75, 121, 135, *144–5*, 145, 153
du Toit, Alexander du 87, *87*
Dubois, Eugene 223
Duckworth, Wynfrid *225*
dunite *106*
Dutton, Clarence 112

Earth
 age 15–20, 33, 40, 45, 52, 58, 186, 208
 atmosphere *See* atmosphere
 chronology 19, 32–3
 circumference 9
 climate 41
 core *See* core
 crust *See* crust
 differentiation 68, 85
 eccentricity 47

elements 39
 geological time 32–3
 geothermal gradient 61, *61*
 magnetic field *See* magnetic field
 mantle *See* mantle
 mass 79
 obliquity 47
 orbit 41, *41*, 47
 precession 47
 reversing polarity 46, *46*
 rotation 47, 48, 62, 81, 94, 209
 shape 9
 snowball Earth 50, *50*
 temperature 61, 104, 209
earth sciences 8
earthquakes 61, 64, 194–5
 Bam 195, *195*
 coseismic interferogram *195*
 earthquake-proof architecture 191, *191*
 Edo *66–7*
 epicenter 73
 Lisbon 70, *70*
 San Francisco 77, *77*, 85
 seismic waves 62, 63, 67, *67*, 78
 subduction 100, *100*
 tsunamis 99, *99*
 uplift 127
Ediacaran biota *54*, 55
Edo earthquake *66–7*
Efremov, Ivan 224–5, *224*, 225
Egyed, László 86
Egyptian civilization 160, *160*
Eichwald, Carl Eduard von *207*
electromagnetic radiation 139
elements 18, 38–9, *38*, *39*
 isotopes 18, 45, 50, 52, 95, 178, 186, 231
 Law of Octaves 38–9
 periodic table 18, 38–9, *39*
Élie de Beaumont, Jean-Baptiste 128, *128*
Enceladus 247, *247*
endolithic organisms 243, *243*
energy 10, 57–9, 84–5, 235
 chemical 84–5
 conversion 84
 elastic 85
 generation from geological resources 161–3
 geothermal 195, 235
 gravitational 85
 heat 58, 61, 84, 85, *85*, 94, 95, 104, 174, 180
 kinetic 84, 97
 nuclear 84–5, 95, *95*, 197, *197*
 plate tectonics 65

radioactive decay 61, 84–5, 95, 104
 solar radiation 97, 209, 235, 243, 245
 subduction zones 64
epicenter 73
Erasmus 172
Eratosthenes 9
Erebus, Mount 98, *98*
erosion 9, *15*, 23
 fluvial *12*, 111–12, 135, 140–1, *140*, *141*, 148–9, *148*, 179
 freeze-thaw process 142, *142*, 179
 glacial 110–11, *111*
 oceanic 114, 124
erratics 111, 130, *130*
estuaries 112, 135
ethane 247
ethene 247
Etna, Mount *32*, 33, *64*, 190, *190*
eukaryotes 244–5
Europa 246–7, *246*
Everest, Mount 101
evolution 18, 33, 48, 131, 135, 205–9, 213, 215, 220, *220*, 223, 226–8, 245
expanding Earth theory 86, 93
extinction events 19, 208, 217, 224, 236–8
Exxon Valdez oil tanker 192

faulting 136–7, *136*, *137*, 194–5, 223
 grabens 137, *137*
 horsts 137
 normal 137
 reverse 137
 rift valleys 89, 223, *223*
 strike-slip 136, 195, 218
 thrust 113, 129, 137
faunal succession 216–17
Figuier, Louis 42
fissures 142, *142*, 153
fjords *111*
flint tools 158–9, 168
Flood narrative 14–15, 20
fluorspar 189
fluvial erosion *12*, 111–12, 135, 140–1, *140*, *141*, 148–9, *148*, 179
folding 121, *121*, 177, 180
 domes 121, *144–5*, 145
Foote, Eunice Newton 200
fossil water 178
fossils 10, 15, 16, 17, 23, 24, 30, 34, *34*, 118, 127, 162, 204–28, *205*, *206*, *210–13*, *216*, *217*, *222*, 226–8, 231, 233, *233*
 cycles in diversity 236–7
 earliest 244, *244*
 faunal succession 216–17
 gaps in record 18, 216–17, 224–5, 236

geological strata 216–17, 223
ichnofossils 238, *239*
Lagerstätten 223, 226
Precambrian *54*, 55
pseudofossils 221, *221*
raindrop fossils 154, *154*
stromatolites 242
Francevillian formation 244, *244*
freeze-thaw process 142, *142*, 179
French, Scott and Romanowicz,
    Barbara A. *102*

Galapagos Islands 218–19, *218*
galena 186
Ganges–Brahmaputra Delta 114, *114*
Ganymede 247
Gauthier-Lafaye, François 95
geological maps 14, *14*, 25, *25*, 30, *31*, 73,
    73, 75, 90, 132, *132*, 134, *134*, 160,
    160, 216
Geological Society of London 132
geological surveying 216
geothermal energy 195, 235
geothermal gradient 61, *61*
Gessner, Conrad 205, *205*
geysers 65, 74–5, *74*, *75*, *96*, 97, 102, 235
Gilbert, Grove Karl 112, *134*, 135, 138
Gilgamesh, Epic of 20, *20*
glaciation 104, 126, 130, 150–1, 175,
    207, 232
    cyclicity 41
    erratics 111, 130, *130*
    glacial erosion 110–11, *111*
    moraine 131, *131*, 150
    sea levels and 142, 219
global warming 71, 163, 178, 200
gneiss 52
Goethe, Johann Wolfgang von 132
gold 160, 184, 199
Gondwanaland 87, *87*, 164
Goosenecks, Utah 150, *150*
gorges 133
grabens 137, *137*
Grand Canyon 36, 37, *37*
granite 58–9, 65, *120*, 143, *147*, 196
Grasberg Mine, Indonesia 184, *184*
gravity 62, 68, 85
    accretion 85
    heat energy 94
    tidal effects 94
Great Dying 236
Great Oxygenation Event 232, 244
Great Rift Valley 223, *223*
Greek civilization 9, 14, 58, 68, 115,
    166, *166*, *167*, 204, *204*
Greenough, George 132, *132*
Guangxi karst landscape 133, *133*

Guettard, Jean-Étienne and Buache,
    Philippe 25, *25*
Gulf War 192, *192*
gypsum *152*, 153, 161, 189

Hadean 55
haematite 242, *242*
Hall, James 110, 122, *122*
Han-Shuo Liu 93, *93*
Harland, W.B. 50, *50*
Haüy, René-Just 119, *119*
Hawaii 75, 102, 218
Hayden, Ferdinand 75
HD 189733b 97
heat 58, 84, 85, *85*, 176
    combustion 174, 180
    convection 61
    Earth's geothermal gradient 61, *61*
    gravitational distortion 94
    metamorphic rocks 170
    phase diagrams 146–7, *146*
    Plutonism 59
    radioactive decay 61, 95, 104
    rock cycle 113, *113*, 120–1
helium 39, 177, *177*
helium-3 199
hematite 153, *153*
Henry Mountains *134*, 135
Herodotus 14
Hesiod 9, 166
Heyn, Björn, et al *103*
Hilgenberg, Ott *86*
Himalayas 81, 90, 101, *101*
Holmes, Arthur 45, 83, *83*
Hooke, Robert 15, 62, 118, 212, *212*
Hopkins, David M. *229*
horsts 137
hotspots 218
human evolution 223, 249
Humboldt, Alexander von 62
Hutton, James 14, 16–17, 28–9, 33,
    58–9, 61, 110, 113, 120–1
    Theory of the Earth 28, 29, 120, *121*
hydrogen 39, 178, 180, 235
hydrothermal vents 164, *234*, 235,
    242, *242*

Ibn Sina (Avicenna) 22, *22*, 109
ice ages 47, *47*, 104, 111, 131, 159
    snowball Earth 50, *50*
Iceland 75, 102, 195
Ichthyosaurus 205
igneous rocks 58–9, 63–5, 88, 110, *113*,
    126, *126*, 143, 219
    batholiths 124, *124*
    oceanic crust 90
    sill 121

Indian Ocean tsunami 99, *99*
indium 188
Industrial Revolution 30, 43, 59, 162, 165,
    173, *173*
Io 94, *94*, 96
iodine 159
ionizing radiation 235
iridescence 122
iridium anomaly 238
iron 90, 97, 104, 160, 166–7, *166*, 173, *173*,
    235, 244
    banded iron formations 232, *232*
iron age 166–7, *166*
isostasy 81, 101, 104
isotopes 18, 45, 50, 52, 95, 178, 186, 231

Jackson, William Henry *74*, 75
Janssonius, Jan *68*
Jupiter 94, 246–7
Jura Mountains 125, *125*
Jurassic 43

karst 133, *133*, 249
Kawah Ijen 56, 57
Kelvin, William Thomson, Lord 18, 40,
    40, 58
kerogen 180
kimberlite 176
Kircher, Athanasius 58, *59*
Klerksdorp spheres 153, *153*
Knutson, Heather 97
Köppen, Wladimir 47
Krakatoa 60, 61
Kuwait, petrochemical pollution 192, *192*

La Oraya mines, Peru 188, *188*
Lagerstätten 223, 226
Lamarck, Jean Baptiste 205–6, 213
Large Low Shear Velocity Provinces 103
Late Heavy Bombardment 50, 242
Laue, Max von 139, *139*
Laurasia 87, 164
lava 59, 109, 179, 190, *190*
    controlling flow 190
    lava lakes 98, *98*, 190
lead 160–1, 186, *187*
Lehmann, Inge
    'Paths Through the Earth' 79, *79*
Lehmann, Johann Gottlob 26, *26*
Lehmann discontinuity 78
Leonardo da Vinci 15, 118, 204–5
    Madonna of the Rocks 23, *23*
    *A Map of the Rivers and Mountains of
        Central Italy* 14
    A Ravine with Waterbirds 7
life *See also* evolution
    endolithic organisms 243, *243*

extraterrestrial 208, 230–1, 240–1, *240*, *241*, 246–7
generally 11, 203–9, 235
hydrothermal vents 164, *234*, 235, 242, *242*
plants 245, *245*
water and 208, 209, 246–8
lignite 174
limestone 133, 161, 173, 249
Linnaeus, Carl 109, 118, *118*
Linth, Arnold Escher von der 129, *129*, 137
Lisbon earthquake 70, *70*
lithium 189, *189*
lithosphere 90
Liu Wentai et al *210*, 211
Llull, Ramon 108, *108*, 204
lobate fault scarps 93
Logan, Alexander 221, *221*
Lower Devonian *34*
lunar eclipse 9
Lyell, Charles *15*, 32–3, 41, 108, 110, 120, 154, 206, 207
Principles of Geology 33, 34, *126*, 127

Macellum of Pozzuoli 32–3, *33*
magma 59, 75
domes 75, 135, *144–5*, 145, 153
hydrothermal vents 235
volcanic rocks 195
magnesium 90, 194, 235
magnetic field 58, *58*, 68, 79, 209
driving force 104
reversing polarity 19, 46, *46*, 88
magnetic resonance imagaging *177*
Mallet, Robert 73, *73*, 90
Malmgren, Finn 142
Manhattan Project 186
mantle *61*, 62, 64, 65, 75, 78–9, 83, 90
asthenosphere 90, 104
convection 83
core-mantle boundary topography 102, 103, *103*
crust thickness 101, 103
density 62, 104
depth 62
hotspots 218
isostasy 81, 101
Large Low Shear Velocity Provinces 103
plate tectonics 62, 64, 65, 82–3
subduction 64, 100, 102, 243
marble 170, *171*
Mars 50, 68, 148–9, *148*, *149*, 207, 240–1, *240*, *241*
Martinique 76
Matthews, Drummond 88

Matuyama, Monotori 46, *46*
meanders 112, 115, *115*, 135, 148, 150, *150*, 159, 170, *170*
Mediterranean Sea 164
Mendeleev, Dmitri 18, 38–9, *39*
Mercury 50, 96, 154, *155*
thermal contraction theory 93, *93*, 128
Mesozoic 27
metals 160–1
mining 173
ore seams 172, 194–5
smelting 160–1, *161*, 162, 164, 166–7, 172, 173
metamorphic rocks 110, *113*, 126, *126*, 170, *171*
metamorphism 110, 176
Meteor Crater, Arizona 138
meteorites 53, 176, 198, 240, 248
methane 96, 97, 149, 232, 235, 241
methanol 247
Michelangelo Buonarroti
Awakening Slave 170, *171*
mid-ocean ridges 63, 64, 81, *85*, 88–90, 164
Milankovitch cycles 47, *47*
Miller, Stanley 208–9, *209*
minerals *See also* crystals
classification 108–9, 116–19, 122, 146
color 122
phase diagrams 146–7, *146*, *147*
mining 30, 161, 162, 164–5, *164*, *165*, *172*, 173–4, *174*, 184, *184*, 188, *188*, 216
off-world 199
slag heaps 164, 188, *188*
soluble metals 189
tailings 184, 188
Mississippi River Valley 115
Mohorovićić discontinuity 78
Mojsisovics, Edmund 35, *35*
molecular compounds 116
monoclines 121
Moon 209
Aitken Basin 198, *198*
craters 48, *49*, 50, *51*, 198, *198*
gravitational effect 94
Mare Crisium 50, *51*
Mare Imbrium 48
minerals on 199
water on 198
Morabito, Linda 94
moraine 131, *131*, 150
Moran, Thomas
The Castle Geyser, Yellowstone 75
The Grand Canyon of the Yellowstone 110

The Great Blue Spring of the Lower Geyser Basin 65
Morley, Lawrence 88
mountains
orogeny 35, 81–2, *82*, *83*, 90, *91*, 112–13, 128–9, *129*, 135
thrust faulting 113, 129, 137
mudflats 135
Murchison object 53, *53*

Natural Bridges National Monument *19*
natural selection 33, 127, 206–7, 218–19
necrolysis 225
Neptune 96
Neptunism 58–9, 122, 143
neutrons 45
Newlands, John 38–9
Newton, Isaac 68
Nicol, William 109
Nietzsche, Friedrich 6
Nile Delta 14
nitrogen 244

oceans 53, 55
acidification 236
continental drift 62–5, 80–1, *80*, *81*, 83
crust 64–5, 88–9, *88*, 104, 231
hydrothermal vents 164, *234*, 235, 242, *242*
mid-ocean ridges 63, 64, 81, *85*, 88–90, 164
oceanic erosion 114, 124
oceanic volcanoes 219
oxygen levels 236
pollution 192, *193*, 201, *201*
sea level 104, 111, 142, 175, 219, 229
spreading oceanic floor 88–90, *88*, *91*
subduction 64, 82, 83, 93, 100, *100*, 243
tidal effects 94
tsunamis 99, *99*, 238
Oklo natural nuclear reactor 95, *95*
Oldham, Richard Dixon 61–2, 78, *78*
opalescence 122
Öpik, Ernst 48
Ordovician 10, 43
orogeny 35, 81–2, *82*, *83*, 90, *91*, 112–13, 128–9, *129*, 135
Osborn, Henry Fairfield *228*
oxbow lakes 115, 148
oxygen 38, 39, 52, 104, 119, 174, 175, 178, 231, 232–3, 235, 236, 242, 243, 244

Pacific Ring of Fire 63
Pacific Superswell region 102, *102*
paleontology 10, 203, 205–28
Paleozoic 27, 87, *87*
Pangaea 62, 68, 80, 92, 164

Pannotia 92
panspermia 208
Pardee, Joseph 141
Patterson, Clair Cameron 186
Pelée, Mount 61, 76
pentagonal network theory 128, *128*
periodic table 18, 38–9, *39*
Permian 43, 162, 175
petrification 225, *225*
petroleum *156–7*, 162, 163, *163*, 180–3, *180–1*
petrochemical pollution 192, *192*, *193*, 201, *201*
Pettit, Don *101*
phase diagrams 146–7, *146*, *147*
Phillips, John 220, *220*
Phlegraean Fields 63, 73
phosphorus 244
photosynthesis 232, 243, 245
Pierce, Stephanie et al *229*
planets
    extrasolar 97, *97*
    formation *84*, 85
    liquid processes 148–9, *148*, *149*
plate tectonics 63–5, 82–3, 88, *88*, 90–3, *91*, 209
    continental crust 90, 175, 223
    mantle 62, 64, 65, 82–3
    mantle convection 83
    mid-ocean ridges 63, 64, 81, *85*, 88–90
    oceanic crust 88–9, 90
    orogeny (mountain-building) 35, 81–2, *82*, 90, *91*, 112–13
    rift valleys 89, 223, *223*
    seismic activity 90
    subduction 64, 82, 83, 93, 100, *100*, 102
    supercontinent cycle 92, *92*
    thermal contraction theory 93, *93*, 128
    thrust faulting 113, 129, 137
Plato 9, 58
Playfair, John 120
Ploieşti oil field 182–3, *182*
Plot, Robert 205, *206*, 215
Plutonism 59, 61, 143
plutonium 231
polarized light photography *106*, 109–10
pollution 133, 163, 183, *183*, 186, 231, 236
    microplastics 201, *201*
    petrochemical 192, *192*, *193*, 201, *201*
Popocatépetl 194, *194*
potash 189
potassium 194
Powell, John Wesley *36*, 37, 111

Precambrian 45, 50, *50*, *54*, 55
pressure 176
    atmospheric 154
    metamorphic rocks 170
    phase diagrams 146–7, *146*
    rock cycle 113, *113*, 120–1
process 10
Proterozoic 55
protoplanetary discs *84*, 85
Psyche 199, *199*
pterodactyl *214*, 215
Pythagoras 9

quartz 97
Quaternary 27

radioactivity/radioactive decay 40, *44*, 45, *45*
    alpha particles 177
    half-life 45
    heat energy 61, 95, 104
    Manhattan Project 186
    nuclear energy 61, 84–5, 95, *95*, 197, *197*
    nuclear waste 197, *197*
    Oklo natural nuclear reactor 95, *95*
    radiometric dating 18–19, 39, 45, 50, 52, 186, 249
Raup, David and Sepkoski, John 236
Read, H. H. 143
Réaumur, René Antoine Ferchault de 109
recurring slope lineae 148–9
reefs 35, 219, *219*
reverse faulting 137
Richat Structure *144–5*, 145
Richthofen, Ferdinand von 35
rift valleys 89, 223, *223*
rivers 114–15, *114*, *115*, 135, 140–1
    braided 112, 135
    deltas 112, 114, *114*, 135, 175
    estuaries 112, 135
    Mars 148–9, *148*
    meanders 112, 115, *115*, 135, 148, 150, *150*, 159, 170, *170*
    oxbow lakes 115, 148
rock cycle 113, *113*, 120–1
Rodinia 92
Romans 161
Röntgen, Wilhelm 139
Rubin, Peter *199*
Rutherford, Ernest 61

Salisbury Crags 120–1
salt 189
Samoa 102
San Andreas Fault 77, 85, 90, 136

San Francisco
    Bay 77, 195
    earthquake 77, *77*, 85
Santorini (Thira) *8*, 9
Saturn 149, 246–7
Scheuchzer, Johann Jakob 211, *211*, 215
Schiaparelli, Giovanni 240
Schmidt, Otto 85
scientific enlightenment 43
sea level 104, 111, 142, 175, 219, 229
Sedgwick, Adam 127, 207
sedimentary rocks 16, 110, *113*, 126, *126*
    deposition 22, 23–4, *24*
    folding 121
    fossils *See* fossils
    strata 16, *16–17*, 26–7, 216–17, 223
    unconformities 28, *28*
sedimentation 114, 205, 216–17, 224, 225, 231
    fossils *See* fossils
seismic activity 73, *73*, 90, 194–5
    coseismic interferogram *195*
    subduction 64, 82, 83, 93, 100, *100*, 102
seismic focus 73
seismic waves 62, 63
    P-waves 78, 79
    S-waves 78
    surface waves 78
    velocity 102, 103, *103*
seismology 73
seismometers 67, *67*
serpentinization 235, 241
Shrewsbury 170, *170*, 173
Siberian Traps 236
Siccar Point, Scotland 29, *29*
silica 143
silicon 39, 52, 230–1
silicon carbide 53, *53*
sill 121
silting 9
Silurian 43
Silurian hypothesis 231
silver 160, 184
sinkholes 133, 238, *238*
slot canyons *12*
smelting 160–1, *161*, 162, 164, 166–7, 172, 173
Smith, William *16–17*, 30, *31*, 205, 213, 216–17, *216*, 217, 224
Snider-Pellegrini, Antonio 62, 80, *80*
Soderblom, Larry 97
sodium 90
soil 109, 113, 114, 159, 194
    mapping 179, *179*
solar system 39, 50, 96–7
    formation 55, *84*, 85

Sprigg, Ray 55
springs 133, 159
Steno (Niels Stensen) 16, 24, *24*, 26, 109, 117, *117*, 205
Steno's Law 117
stone age 166
stone tools 158–9, 168, *168*
Stonehenge *158*
Strabo 161
strata 16–17, 23–4, *24*, 26–30, *31*, 57, 126
　faunal succession 216–17, 223
　Grand Canyon 36, 37, *37*
　magnetization 46, *46*
　Smith's survey of *16–17*, 216–17
　stratigraphy 126, 205, 213, 216–17, 238
stromatolites 242
subduction 64, 82, 83, 93, *100*, 102, 243
subsidence 34
Sun 18
　gravitational effect 94
　solar radiation 47, 97, 209, 235, 243, 245
supercontinent cycle 92, *92*
Sylbaris, Ludger 76, *76*
synclines 121

Taller Mecanico formation 154, *154*
talus 124
Tambora, Mount 61, 71, *71*
taphonomy 224–5
Tertiary 27
Tethys Ocean 164
Tharp, Marie 89, *89*
thermal contraction theory 93, *93*, 128
thermodynamics 40, 58
Thomsen, Christian Jürgensen 166
thrust faulting 113, 129, 137
Tibetan Plateau 101, *101*
tidal effects 94
tillites 50, *50*
time
　age of the Earth 15–20, 33, 40, 45, 52, 58, 186, 208
　chemical elements 39
　crater counting 48, *49*, 50
　cycles of geological activity 15–19, 32–3, *33*
　deposition 14
　erosion *15*
　faunal succession 216–17, 220, *220*
　geological time 27–9, *27*, 32–3, 43, *43*, 120, 126
　measurement 10, 13, 14–19, 39
　oldest rocks 52–3, *52*, *53*
　radiometric dating 18–19, 39, 45, 50, 52, 186, 249

thermodynamics 40, 58
three-age system 166–7
tin 165, *165*, 166, 167
Titan 149, *149*
tomography *103*, 166
Triassic 43
Triton 96–7, *96*
tsunamis 70, 99, *99*, 238
Tunguska Event 138, *138*
tunneling 185, *185*, 196–7, *196*, *197*
Turin Papyrus 160, *160*
Turner, J.M.W.
　Sunset 71, *71*
Tuzo Wilson, John 90, *91*, 92, *92*

U-shaped valleys 111, 131
unconformities 28, *28*
uniformitarianism 16–17, 18, 19, 33, 126, 127
uplift 29, 34–5, 37, 112–13, *113*, 121, 127, 135, 150, 164
uranium *44*, 45, *45*, 50, 52, 95, 177, 186, 231, 249
Urey, Harold 208
Ussher, Archbishop James 15

V-shaped valleys 112, 135, 140
veins 120, *120*
Vesuvius 63, *73*
Vine, Frederick 88, *88*
volatiles 154
volcanoes 9, *42*, 58–61, *58*, *59*, 126
　ash 194, *194*
　Ball's Pyramid 124, *124*
　caldera 32, *63*, 75
　cataclysmic 19
　geysers *65*, 74–5, *74*, *75*, 96, 97
　islands *8*, 9
　largest recorded explosion 71, *71*
　lava *56*, 59, 179, 190, *190*, 194
　lava lakes 98, *98*
　minerals associated with 194–5
　oceanic 219
　plate tectonics 63–5
　plumes *65*, 71, 75, 94, *94*, *96*, 97, 102, *102*, *103*, 240, 247
　rock cycle *113*
　soils 179, 194
　subduction 64, 82, 83, 100, *100*
　supervolcanos *65*, 74–5
　tsunamis 70
　volcanic rock 27

Walcott, Charles 228
Wallace, Alfred Russel 33, 127, 207
WASP-76b 97
water *See also* glaciation; oceans; rivers

acidification 133, 136
aquifers 178
fluvial erosion *12*, 111–12, 135, 140–1, *140*, *141*, 148–9, *148*, 179
　generally 52–3, 246–8
　life and 208, 209, 246–8
Moon 198
Neptunism 58–9
petrification 225, *225*
sea level 104, 111, 142, 175, 219, 229
tetragonal ice 247
water table 195
weathering 113
Wedgwood scale 122, *122*
Wegener, Alfred 47, 62, 80–1, *81*, 87, 94
Werner, Abraham Gottlob 58–9, 122, *123*, 143
Wheal Coates tin mine 165, *165*
Williams, William
　The Iron Bridge 173
　Morning View of Coalbrookdale 162
Wilson Cycle 92, *92*
Wutky, Michael
　The Phlegraean Fields 63

X-ray crystallography 139, *139*
Xenophanes 204

Yaggy, Levi Walter 43, *43*
Yellowstone *65*, 74–5, *74*, *75*, 102, *110*, 218, *235*

Zhang Heng *67*
zinc 166
zircon 52, 248, *248*

Picture Credits

Images on the pages listed below are reproduced by kind permission of the owners. The publishers have made every effort to trace copyright holders and to obtain their permission for the use of copyright material. The publishers apologize for any errors or omissions and will gratefully incorporate any corrections in future reprints if notified.

Specific acknowledgments are as follows: l=left, r=right, t=top, c=center, b=bottom.
2 Public domain, via Wikimedia Commons; **7** Gravure Francaise/Alamy; **8** borchee/iStockphoto; **11** David Bainbridge; **12–13** David Bainbridge; **14** Svintage Archive/Alamy; **15** Archivac/Alamy; **16–17** Reproduced by permission of the Geological Society of London; **19** David Bainbridge; **20** Public domain, via Wikimedia Commons (Photographer: U0045269/CC BY-SA 4.0); **21** The Picture Art Collection/Alamy; **22** Public domain; **23** Public domain, via Wikimedia Commons; **24** The Print Collector/Alamy; **25** Public domain, via Wikimedia Commons; **26** Linda Hall Library (www.lindahall.org) (CC BY 4.0); **27** The History Collection/Alamy; **28** The Natural History Museum/Alamy; **29** David Bainbridge; **31** Classic Collection/Alamy; **32** David Bainbridge; **33** Public domain, via Wikimedia Commons; **34** Natural History Museum, London (CC0-1.0); **35** Public domain; **36** Science History Images/Alamy; **37** David Bainbridge; **38** Science History Images/Alamy; **40** Chronicle/Alamy; **41** Linda Hall Library (www.lindahall.org) (CC BY 4.0); **42** Chronicle/Alamy; **43** Public domain, via Wikimedia Commons; **44** Ted Kinsman/Science Photo Library; **45** Ted Kinsman/Science Photo Library; **46** Public domain; **47** Linda Hall Library (www.lindahall.org) (CC BY 4.0); **49** NASA; **50** From: Harland, W.B. 'Critical evidence for a great infra-Cambrian glaciation'. *Geol Rundsch* 54, 45-61; **51** World History Archive/Alamy; **52** Smithsonian Institution (NMNH - Mineral Sciences Dept./NMNH 116542-1/CC0); **53** From: 'Lifetimes of interstellar dust from cosmic ray exposure ages of presolar silicon carbide', Philipp R. Heck et al, January 13, 2020/117 (4) 1884-1889/ https://doi.org/10.1073/pnas.1904573117; **54** Ilya Bobrovskiy; **55** J Marshall – Tribaleye Images/Alamy; **56–57** Westend61 GmbH/Alamy; **58** Public domain, via Wikimedia Commons; **59** Chronicle/Alamy; **60** Volgi Archive/Alamy; **62** Vladirina32/Shutterstock; **63** Logic Images/Alamy; **64** The eruption of Mount Etna in 1669. Wellcome Collection (Public domain); **65** Historic Collection/Alamy; **66–67** Art Collection 3/Alamy; **67** Science and Society Picture Library/Getty Images; **68t** GRANGER – Historical Picture Archive/Alamy; **68c** The Granger Collection/Alamy; **69** Public domain, via Wikimedia Commons; **70** The Picture Art Collection/Alamy; **71** Public domain, via Wikimedia Commons; **72** Linda Hall Library (www.lindahall.org) (CC BY 4.0); **73** Royal Astronomical Society/Science Photo Library; **74** Natural History Library/Alamy; **75** Alpha Stock/Alamy; **76** ART Collection/Alamy; **77** Everett Collection Inc/Alamy; **78** Public domain, via Wikimedia Commons; **79l** Public domain; **79r** Public domain; **80l** Science History Images/Alamy; **80r** Science History Images/Alamy; **81** The Natural History Museum/Alamy; **82** Courtesy of Dr Christoph Hauser; **83** Author's collection; **84** Science History Images/Alamy; **85** Creator: Fabio Crameri (Original version: 25.10.2021/This version: 10.08.2023/CC BY-SA 4.0. This graphic by Fabio Crameri based on data by J. H. Davies (2013) is available via the open-access s-ink.org repository); **86** Public domain, via Wikimedia Commons; **87** Public domain; **88** Bookend/Alamy; **89** GRANGER – Historical Picture Archive/Alamy; **91** From: 'Evidence from ocean islands suggesting movement in the Earth', J. Tuzo Wilson (Published by The Royal Society: 28 October 1965/https://doi.org/10.1098/rsta.1965.0029); **92** From: 'Fifty years of the Wilson Cycle Concept in Plate Tectonics: An Overview', R. Wilson, et al, Geological Society, London, Special Publications (vol. 470, pp. 1–17, June 2019/https://doi.org/10.1144/sp470-2019-58) (CC BY 4.0); **93** From: 'Thermal Contraction of Mercury', Han-Shou Liu, Goddard Space Flight Center, NASA, Washington, D.C./https://ntrs.nasa.gov/api/citations/19700022338/ downloads/19700022338.pdf; **94** NASA Image Collection/Alamy; **95** François Gauthier-Lafaye; **96** Stocktrek Images, Inc./Alamy; **97** NASA/JPL-Caltech/H. Knutson (Harvard-Smithsonian CfA), Public domain via Wikimedia Commons; **98** Zuma Press, Inc./Alamy; **99** Science History Images/Alamy; **100** From: Fig. 4. B. Bouyahiaoui et al. 'Crustal structure of the eastern Algerian continental margin and adjacent deep basin: implications for late Cenozoic geodynamic evolution of the western Mediterranean.' *Geophysical Journal International* (201) 3 (2015): 1912-1938. Reproduced by permission of Oxford University Press on behalf of the Royal Astronomical Society; **101** NASA (Photograph taken from the International Space Station in May 2012 by astronaut Don Pettit); **102** © Dr Scott French/UC Berkeley; **103** From: 'Core-mantle boundary topography and its relation to the viscosity structure of the lowermost mantle,' Björn H. Heyn, Clinton P. Conrad, Reidar G. Trønnes, 1 June 2020/https://doi.org/10.1016/j.epsl.2020.116358 (CC BY 4.0); **104** Mike Beauregard from Nunavut Canada, via Wikipedia (CC BY 2.0); **105** Photo: ©ESA. Credit: Julien Aubert, IPGP/CNRS/CNRS Photothèque/(ESA Standard License); **106** Microckscopia/Science Photo Library; **108** Album/Alamy; **109** Garitan, via Wikipedia (CC BY-SA 4.0); **110** ARTGEN/Alamy; **111** Tatiana Popova/Shutterstock; **112** David Bainbridge; **114** ©ESA (CC BY-SA 3.0 IGO); **115** EnesTaha/Adobe Stock; **116** GRANGER – Historical Picture Archive/Alamy; **117** Public domain; **118** Book Worm/Alamy; **119** Public domain; **120** © The Hunterian, University of Glasgow; **121** Royal Institution of Great Britain/Science Photo Library; **122** Linda Hall Library (www.lindahall.org) (CC BY 4.0); **123** Linda Hall Library (www.lindahall.org) (CC BY 4.0); **124** Blue Planet Archive LLC/Alamy; **125** Hemis/Alamy; **126** Jimlop Collection/Alamy; **127** Julian Cartwright/Alamy; **128** Bibliothèque Mines Paris – PSL; **129** ETH-Bibliothek Zürich (Public domain); **130** Penta Springs Limited/Alamy; **131** The Picture Art Collection/Alamy; **132** Public domain, via Wikimedia Commons; **133** Michael Thomas/Alamy; **134** Linda Hall Library (www.lindahall.org) (CC BY 4.0); **135** NASA (December 24, 2019, provided by the ISS Crew Earth Observations Facility and the Earth Science and Remote Sensing Unit, Johnson Space Center. The image was taken by a member of the Expedition 61 crew); **136** Public domain; **137** © Mohammad Goudarzi (CC BY 3.0); **138** World History Archive/Alamy; **139** Public domain, via Wikimedia Commons; **140** David Bainbridge; **141** David Bainbridge; **142** Nature Picture Library/Alamy; **143** Frontispiece of *The Granite Controversy* by H. H. Read, Interscience, New York, drawn by D. A. Walton. **144–145** Landsat 7, USGS, NASA; **146** T.H. Green; **147** From: 'Effect of anorthite on granite phase relations: Experimental data and models', Michel Pichavant, Catherine Weber, Arnaud Villaros, 2019. https://doi.org/10.1016/j.crte.2019.10.001 (CC BY 4.0); **148** NASA Image Collection/Alamy; **149t** NASA/JPL-Caltech/ASI/USGS; **149b** NASA/JPL-Caltech/MSSS/JHU-APL; **150** David Parker/Science Photo Library; **151** Martin Bond/Science Photo Library; **152** Javier Trueba/Science Photo Library; **153** © Klerksdorp Museum; **154** Verisimilus, via Wikipedia (CC BY-SA 3.0); **155** NASA/Johns Hopkins University Applied Physics Laboratory/Carnegie Institution for Science; **156–157** Jeeraphun Juntree/Alamy; **158** Laurence Berger/Shutterstock; **159** Gainew Gallery/Alamy; **160** Skimage/Alamy; **161** The History Collection/Alamy; **162** ARTGEN/Alamy; **163** Public domain, via Wikimedia Commons; **164** Anna Kucherova/Shutterstock; **165** eye35/Alamy; **166** The Fitzwilliam Museum/Bridgeman Images; **167** Witold Skrypczak/Alamy; **168** Nationalmuseet, via Wikimedia Commons (CC BY SA-4.0); **169** Cynthia Lee/Alamy; **170** Paul White – UK Cities/Alamy; **171** Bridgeman Images; **172** Pictorial Press Ltd/Alamy; **173** Public domain, via Wikimedia Commons; **174** History and Art Collection/Alamy; **175** Ron Blakey ©2016 Colorado Plateau Geosystems Inc.; **176** The Natural History Museum/Alamy; **177** toubibe/Pixabay; **178** Kara Grubis/Shutterstock; **179** Public domain; **180–181** Sueddeutsche Zeitung Photo/Alamy; **182** Niday Picture Library/Alamy; **183** Monty Fresco/Hulton Archive/Getty Images; **184** akhid7790/Shutterstock; **185** Zhihua Feng/Alamy; **188** imageBROKER.com/Alamy; **189** FREEDOM_WANTED/Alamy; **190** Courtesy of L. Villari, Istituto Nazionale di Geofisica e Vulcanologia (INGV) (CC BY 4.0); **191** Morio, via Wikipedia (CC BY-SA 3.0); **192** Public domain, via Wikimedia Commons (Tech. Sgt. David McLeod); **193** PA Images/Alamy; **194** Byelikova/Dreamstime; **195** © David Small (UZH) University of Zürich, Switzerland; **196** NB/ROD/Alamy; **197** Benjamin Suomela/Yle; **198** NASA/GSFC/University of Arizona; **199** Zuma Press, Inc./Alamy; **200** From: Ben Callow, Ismael Falcon-Suarez, Sharif Ahmed, Juerg Matter, 'Assessing the carbon sequestration potential of basalt using X-ray micro-CT and rock mechanics', International Journal of Greenhouse Gas Control, Volume 70, 2018, pp. 146-156, ISSN 1750-5836, https://doi.org/10.1016/j.ijggc.2017.12.008; **201** © Dr. Victoria M. Fulfer; **202–203** Puwadol Jaturawutthichai/Alamy; **204** Mark Landon, via Wikimedia Commons (CC BY 4.0); **205** Linda Hall Library (www.lindahall.org) (CC BY 4.0); **206** Linda Hall Library (www.lindahall.org) (CC BY 4.0); **207** From: 'Arbor vitae animalis' [Tree of life of the animals]. Karl Eduard von Eichwald, Zoologia (Vilnae: J. Zawadzki, 1829), Vol. I, between p. 40 and p. 41. Public domain; **209** Bettmann/Getty Images; **210** Chinese Materia Medica illustration, Ming: Dragon Bone. Wellcome Collection. Public domain; **211** Ghedoghedo, via Wikimedia Commons (CC BY-SA 4.0), **212** From the British Library archive, via Bridgeman Images; **213** Public domain; **214–215** Linda Hall Library (www.lindahall.org) (CC BY 4.0); **215** Public domain, via Wikimedia Commons; **216** Historic Collection/Alamy; **217** Public domain, via Wikimedia Commons; **218** *Jacques Descloitres, MODIS Land Rapid Response Team,* NASA/GSFC, 2002; **219** Science History Images/Alamy; **220** Public domain; **221** The Book Worm/Alamy; **222** H. Raab (User: Vesta), via Wikipedia (CC BY-SA 3.0); **223** Joanna Rigby-Jones/Shutterstock; **224** Public domain; **225** Public domain; **226** With permission of ROM (Royal Ontario Museum), Toronto, Canada. Photo: Jean-Bernard Caron. © ROM; **227** 3D Reconstruction and Image © Stephanie Pierce. From: Pierce, S.E., Clack, J.A. and Hutchinson, J.R., 2012: 'Three-dimensional limb joint mobility in the early tetrapod Ichthyostega', *Nature*, 486 (7404), pp. 523–526; **228** From Osborn, Henry Fairfield, 1929, 'The titanotheres of ancient Wyoming, Dakota, and Nebraska'. Publisher: Dept. of the Interior, U.S. Geological Survey, volume 55, plate XLI. Biodiversity Heritage Library: https://www.biodiversitylibrary.org/page/26854372; **229** Reproduced from David M. Hopkins, 'Cenozoic History of the Bering Land Bridge.' *Science*, Vol 129, No. 3362 (June 5, 1959): pp. 1519–28. Published by: American Association for the Advancement of Science. http://www.jstor.org/stable/1757656; **230** Everett Collection Inc/Alamy; **231** AndreaScala/Getty Images; **232** Dirk Wiersma/Science Photo Library; **233** Didier Descouens, via Wikimedia Commons (CC BY-SA 4.0); **234**-235 Galih/Alamy; **235** David Bainbridge; **236** Reproduced from David M. Raup et al., 'Mass Extinctions in the Marine Fossil Record', *Science*, Mar 19 1982, Vol 215 (Issue 4539): pp. 1501-3. DOI: 10.1126/science.215.4539.1501, AAAS. © 1982 American Association for the Advancement of Science; **237** From: Rohde, R., Muller, R. 'Cycles in fossil diversity'. *Nature* 434, 208–210 (2005). https://doi.org/10.1038/nature03339; **238** Sergey Novikov/Shutterstock; **239** Arterra Picture Library/Alamy; **240** NASA, via Wikipedia; **241** NASA/JPL-Caltech/MSSS; **242** © Dominic Papineau; **243** From: Hidalgo-Arias, Andrea, Víctor Muñoz-Hisado, Pilar Valles, Adelina Geyer, Eva Garcia-Lopez, and Cristina Cid. 2023. 'Adaptation of the Endolithic Biome in Antarctic Volcanic Rocks', *International Journal of Molecular Sciences* 24, no. 18: 13824. (Open access CC license.) https://doi.org/10.3390/ijms241813824; **244** From: © El Albani A, Bengtson S, Canfield DE, Riboulleau A, Rollion Bard C, Macchiarelli R, et al. (2014). 'The 2.1 Ga Old Francevillian Biota: Biogenicity, Taphonomy and Biodiversity'. PLoS ONE 9(6): e99438. https://doi.org/10.1371/journal.pone.0099438 (CC BY 4.0); **245** Images from article 'A one-billion-year-old multicellular chlorophyte.' Tang et al. (2020); **246** NASA:JPL-Caltech:SETI Institute; **247** NASA Image Collection/Alamy; **248** © John W. Valley, University of Wisconsin; **249** Maxime Aubert, Griffith University.